UNIFIED S-BAND TELECOMMUNICATION TECHNIQUES FOR APOLLO

VOLUME II

MATHEMATICAL MODELS AND ANALYSIS

By John H. Painter and George Hondros

Manned Spacecraft Center
Houston, Texas

NATIONAL AERONAUTICS AND SPACE ADMINISTRATION

ABSTRACT

This is the second volume in a series done at Manned Spacecraft Center, documenting communication techniques used in the Apollo Unified S-band Telecommunications and Tracking System. As stated in the first volume, NASA TN D-2208, the present document is concerned with detailed mathematical modeling of certain channels in the system. Specifically, this volume provides simple mathematical tools usable for predicting, approximately, the performance of various communications and tracking channels in the system.

FOREWORD

The content of this publication was originally published as NASA Apollo Working Paper No. 1184, and was entitled "Unified S-Band Telecommunication Techniques for Apollo, Volume II, Mathematical Models and Analysis." Volume I of this series, by the same authors, is entitled "Functional Description." This publication, which has become Volume II, is nearly identical to the Working Paper mentioned above, incorporating only a few minor changes in the form.

The authors wish to acknowledge the time and effort of members of the technical staffs of Jet Propulsion Laboratory, Pasadena, California, and Motorola Military Electronics Division, Western Center, Scottsdale, Arizona. The first draft was reviewed by the following individuals from Jet Propulsion Laboratory, each performing a separate review function:

> M. Koerner
> L. Couvillon
> R. Titsworth
> M. Brockman
> R. Toukdarian
> W. Victor

The following individuals from Motorola, Inc. reviewed the first draft in its entirety:

> Dr. S. C. Gupta
> T. G. Hall

A word of thanks is due to Messrs. B. D. Martin and M. Easterling of JPL for continued assistance and inspiration during the preparation of this volume.

The authors wish to dedicate this publication to G. Barry Graves who made the writing of this series of documents possible, and to Ted Freeman, one of those individuals whom this work was intended to benefit.

CONTENTS

Section			Page
	SUMMARY		1
	SYMBOLS		2
1.0	INTRODUCTION		5
	1.1	Background and Purpose of the Document	5
	1.2	Theoretical Approach	5
		1.2.1 Analytical Scope	5
		1.2.2 Method of Presentation	6
		1.2.3 Assumptions	6
	1.3	System Description	7
2.0	GROUND-TO-SPACECRAFT CHANNEL ANALYSES		7
	2.1	Carrier Tracking Channel	7
	2.2	The Voice Channel	8
	2.3	The Up-Data Channel	11
3.0	SPACECRAFT-TO-GROUND PHASE MODULATED CHANNEL ANALYSES		13
	3.1	Carrier Tracking Channel	13
	3.2	Angle Tracking Channel	15
	3.3	Ranging Channel	15
		3.3.1 Clock Loop Threshold	16
		3.3.2 Range Code Acquisition Time	18
	3.4	The PCM Telemetry Channel	19
	3.5	The Voice Channel	22
	3.6	Biomedical Data Channels	25
	3.7	The Emergency Voice Channel	28
	3.8	The Emergency Key Channel	31

Section	Page
4.0 SPACECRAFT-TO-GROUND FREQUENCY MODULATED CHANNEL ANALYSES	33
4.1 Carrier Demodulation Channel	33
4.2 Television Channel	35
4.3 PCM Telemetry Channel	37
4.4 Voice Channel	40
4.5 Biomedical Data Channel	42
APPENDIX A — ANGLE MODULATION	A-1
A.1 Basic Considerations	A-1
A.2 The Carrier with K Subcarriers	A-4
A.3 The Carrier with K Subcarriers and Range Code	A-6
APPENDIX B — NOISE	B-1
B.1 The Narrow-Band Gaussian Random Process	B-1
B.2 Angle Modulated Carrier Plus Noise	B-3
B.3 Transmission of Signal Plus Noise Through a Perfect Band-pass Limiter	B-5
B.4 Transmission of Signal Plus Noise Through a Perfect Product Device	B-6
B.4.1 A Nonprelimited Product Detector	B-6
B.4.2 A Prelimited Product Detector	B-7
B.4.3 A Nonprelimited Product Mixer	B-9
APPENDIX C — PHASE-LOCKED LOOP THEORY	C-1
C.1 A Physical Approach to the Phase-locked Loop	C-1
C.2 The Linearized Model of the Phase-locked Loop	C-4
C.2.1 The Closed Loop Transfer Functions	C-5
C.2.2 Modulation Tracking Error	C-9
C.2.3 Loop Phase Noise	C-13
C.2.4 Threshold Prediction	C-15

Section	Page
C.3 Signal and Noise Characteristics of Prelimited Phase-Locked Loops	C-18
C.3.1 Limiter Effects on Loop Parameters	C-19
C.4 Modulation Restrictive Loop	C-20
C.4.1 Loop Noise Bandwidth Above Threshold	C-22
C.5 Prefiltered Modulation Tracking Loops	C-24

APPENDIX D — PRODUCT DEMODULATION D-1

    D.1 Linear Product Demodulator D-1

        D.1.1 Detection of Sinusoidal Subcarriers D-2
        D.1.2 Detection of Arbitrary Baseband Modulation D-4
        D.1.3 Noise Characteristics D-6
        D.1.4 Output Signal-to-noise Ratios D-6

            D.1.4.1 Subcarrier and band-pass filter D-6
            D.1.4.2 Baseband modulation and low-pass filter D-7

    D.2 Prelimited Product Demodulators D-9

APPENDIX E — DEMODULATION WITH MODULATION TRACKING LOOPS E-1

    E.1 Detection of Sinusoidal Subcarriers and Arbitrary Baseband Modulation E-3

    E.2 Noise Characteristics E-4

        E.2.1 Low-pass Output Filter E-5
        E.2.2 Band-pass Output Filter E-6

    E.3 Output Signal-to-noise Ratios E-8

        E.3.1 Subcarrier and Band-pass Filter E-8
        E.3.2 Baseband Modulation and Low-pass Filter . . . E-10

APPENDIX F — SPECIALIZED DETECTORS F-1

    F.1 Range Clock Receiver and Code Correlator F-1

        F.1.1 Signal Treatment F-2
        F.1.2 Noise Treatment F-5

Section	Page
F.1.3 Signal-to-noise Ratios	F-10
F.1.4 Receiver Threshold	F-12
F.1.5 Range Code Acquisition Time	F-13
F.2 PCM Telemetry Subcarrier Demodulator	F-18
F.2.1 Output Data Treatment	F-19
F.2.2 Reference Loop Treatment	F-20
F.3 The Residual Carrier Tracking Receiver (Ground)	F-22
F.4 The Residual Carrier Tracking Receiver (Spacecraft)	F-25
F.5 The Spacecraft Turnaround Ranging Channel	F-29
F.5.1 Equivalent Noise	F-32
F.5.2 Equivalent Signal	F-35
APPENDIX G — PHASE MODULATED SIGNAL DESIGN	G-1
G.1 Solution for Modulation Indices	G-2
G.2 Maximization of Subcarrier Channel Signal-to-noise Ratios	G-7
G.3 Boundary Condition on Residual Carrier	G-7
G.4 Signal Efficiency	G-9
APPENDIX H — SUPPLEMENTARY THEORY	H-1
H.1 The Equivalent Noise Bandwidth of Linear Networks	H-1
H.2 Equivalent Noise Temperature of Linear Systems	H-5
H.2.1 Single Networks	H-5
H.2.2 Cascaded Networks	H-7
H.3 The Band-pass Amplitude Limiter	H-10
H.4 The Range Equation	H-13
H.5 Antenna Polarization Loss	H-17
H.6 Intelligibility of Clipped Voice	H-19

Section	Page
REFERENCES	R-1

TABLES

Table		Page
C.2.2-I	INPUT FUNCTIONS	C-10
C.2.4-I	CONFIDENCE VALUES VERSUS LOSS-LOCK PROBABILITIES	C-17
C.4-I	INPUT SNR VERSUS LOSS-LOCK PROBABILITIES	C-22
C.5-I	INPUT SNR VERSUS LOSS-LOCK PROBABILITIES	C-26
F.1.1-I	PROGRAM STATE VERSUS CORRELATION	F-4

FIGURES

Figure		Page
2.2-1	Up-link channels	9
3.3-1	The ranging channel	17
3.4-1	PCM telemetry channel	20
3.5-1	The voice channel	23
3.6-1	The biomedical data channel	26
3.7-1	The emergency voice channel	29
3.8-1	The emergency key channel	32
4.1-1	The FM carrier channel	34
4.2-1	The television channel	36
4.3-1	PCM telemetry channel	38
4.4-1	The voice channel	41
4.5-1	The biomedical data channel	43
B.1-1	Input noise spectrum	B-2
B.3-1	Limiter model	B-5
B.4.1-1	Nonprelimited product detector	B-6
B.4.2-1	Prelimited product detector	B-8
B.4.3-1	Nonprelimited product mixer	B-9
C.1-1	Physical loop model	C-1
C.2-1	Linear loop model	C-5
C.2.1-1	Asymptotic Bode plots of transfer functions	C-8
C.3-1	Prelimited phase-locked loop	C-18
D.1-1	Demodulator configuration	D-1
D.2-1	Demodulator configuration	D-9

Figure		Page
E-1	Demodulator configuration	E-1
E-2	Asymptotic Bode plot	E-2
F.1-1	Range clock receiver	F-1
F.1.5-1	Error probability versus signal-to-noise density ratio	F-17
F.2-1	PCM telemetry subcarrier demodulator	F-18
F.3-1	Carrier tracking receiver	F-22
F.4-1	Carrier tracking receiver	F-25
F.5-1	Spacecraft turnaround channel	F-29
H.1-1	Linear network model	H-1
H.1-2	Contour of integration	H-4
H.2.1-1	Equivalent noise temperature of a noisy linear network	H-6
H.2.2-1	Cascaded linear noisy networks	H-8
H.2.2-2	Cascaded passive and noisy networks	H-10
H.3-1	Band-pass limiter model	H-11
H.3-2	Limiter signal and noise suppression versus input SNR	H-12
H.3-3	Exact and approximate signal suppression	H-13
H.4-1	Communication link model	H-14
H.4-2	Antenna geometry	H-14
H.5-1	Elliptic polarization	H-18

UNIFIED S-BAND TELECOMMUNICATION TECHNIQUES FOR APOLLO

VOLUME II

MATHEMATICAL MODELS AND ANALYSIS

By John H. Painter and George Hondros
Manned Spacecraft Center

SUMMARY

This is the second volume in a series done at Manned Spacecraft Center, documenting communication techniques used in the Apollo Unified S-band Telecommunications and Tracking System. As stated in the first volume, NASA TN D-2208, the present document is concerned with detailed mathematical modeling of certain channels in the system. Specifically, this volume provides simple mathematical tools usable for predicting, approximately, the performance of various communications and tracking channels in the system.

SYMBOLS

Signal structure:

A	peak amplitude of a sinusoidal carrier
$c(t)$	a square waveform, having values +1 and -1, which may be subscripted for identification
$f_b(t)$	baseband modulation function of a frequency-modulated carrier
$f(t)$	arbitrary signal function
$l(t)$	output function of an ideal bandpass limiter
$m(t)$	output function of an ideal multiplier
$s(t)$	a desired signal time function which may appear with identifying subscripts
$\Delta\omega_i$	peak frequency deviation in radians/sec of a sinusoidal subcarrier on a down-link frequency modulated carrier
$\Delta\varphi_i$	peak phase deviation, in radians, of a sinusoidal subcarrier of a down-link phase modulated carrier
$\Delta\varphi_j$	peak phase deviation, in radians, of a sinusoidal subcarrier of an up-link phase modulated carrier
$\Delta\varphi_r$	peak phase deviation in radians of a pseudo-random ranging code on an up-link phase modulated carrier
$\Delta\varphi_r\text{eff}$	effective peak phase deviation of a pseudo-random ranging code on down-link phase modulated carrier
$\theta_i(t)$	equivalent phase modulation function on a subcarrier on a down-link frequency modulated carrier
$\varphi_i(t)$	equivalent phase modulation function on a down-link subcarrier
$\varphi_j(t)$	equivalent phase modulation function on an up-link subcarrier
$\varphi_s(t)$	equivalent phase modulation function of an angle modulated carrier

$\psi(t)$	a modulating function of an angle modulated sinusoidal carrier
ω_c	unmodulated radian frequency of a sinusoidal carrier
ω_i	radian frequency of a down-link subcarrier
ω_j	radian frequency of an up-link subcarrier

Phase-locked loops:

A	amplitude of input sinusoid
A_V	amplitude of VCO sinusoid
B_N	equivalent one-sided closed-loop noise bandwidth
$E(s)$	Laplace transform of $e(t)$
e_m	maximum value of $e(t)$
$e(t)$	loop modulation tracking error function
$F_L(s)$	Laplace transform of $h(t)$
$G(s)$	equivalent closed loop input output transfer function
$h(t)$	loop filter impulse response function
K	open loop gain constant
K_V	VCO constant
p	pole frequency of the loop filter
$V_D(s)$	Laplace transform of $v_d(t)$
$v_d(t)$	VCO driving function
$v_v(t)$	output function of voltage controlled oscillator (VCO)
x	peak factor for VCO phase jitter
z	zero frequency of the loop filter
ξ	loop damping ratio
σ_φ^2	variance of VCO phase jitter process
$\Phi_i(s)$	Laplace transform of $\varphi_i(t)$ in complex variable s

$\Phi_o(s)$	Laplace transform of output phase function
$\varphi_i(t)$	input phase function
$\varphi_o(t)$	VCO output phase function
ω_n	loop natural resonant frequency

Noise:

$n(t)$	a sample function of a narrow-band-limited white Gaussian noise process
$x(t), y(t)$	sample functions of independent low frequency white Gaussian noise processes derived from $n(t)$
$\sigma_x^2, \sigma_y^2, \sigma_n^2$	variances of the variables n, x, y
$\|\Phi_x\|, \|\Phi_y\|, \|\Phi_n\|$	absolute, nonzero, values of the flat spectral densities
$\Phi_x(\omega), \Phi_y(\omega), \Phi_n(\omega)$	noise spectral densities of the functions $x(t), y(t), n(t)$
$\|\Phi_\varphi\|$	absolute, nonzero, value of the flat spectral density
$\Phi_\varphi(\omega)$	spectral density of an equivalent low frequency white Gaussian phase process derived from $n(t)$

Miscellaneous:

B	filter transmission bandwidth
B_L	equivalent square transmission bandwidth of an ideal bandpass limiter
B_o	bandwidth of an output filter
f_m	midfrequency of an output bandpass filter
K_f	filter transmission constant
K_p	peak to rms factor
K'_φ	multiplier constant of an ideal multiplier
M	frequency multiplication ratio
N_o	an output noise power
P_L	output power of an ideal bandpass limiter

$\frac{S}{N}$, SNR   signal-to-noise power ratio

V_L voltage limiting level of an ideal bandpass limiter

α_L signal suppression factor of an ideal bandpass limiter

1.0 INTRODUCTION

1.1 Background and Purpose of the Document

Work similar to this volume has been performed previously, external to Manned Spacecraft Center. In general, such work was fragmentary and was performed to meet certain immediate needs such as responses to NASA requests for proposal. When work began on this volume in mid-1963, the authors felt that a need existed for a comprehensive tutorial document setting down general analyses of the types of channels employed in the Apollo system. It was felt that for such a document to be useful to NASA engineers it should contain, in appendix form, sufficient basic explanation to completely and independently support the body of the document. This work is the authors' answer to that need.

1.2 Theoretical Approach

1.2.1 Analytical Scope

The analysis presented in this document has been performed with the aim of obtaining tractable equations with which the output data quality can be predicted for each channel for a variety of transmission modes and conditions.

The approach has been to derive output data signal-to-noise power ratios which are related to the input carrier-to-noise ratio for each communication channel, by an expression containing generalized signal modulation parameters and channel transmission parameters. The requirement that the channel equations be tractable was taken to imply that the expressions be relatively simple and amenable to hand calculation with the aid of mathematical tables. Additionally, it was desired that the form of the channel equations should give some intuitive insight into the operation of the channel.

Input-output signal-to-noise ratio relations were derived separately for each type demodulator and each type signal. Where simplifying assumptions were made, they were stated explicitly in the derivations. Additionally, the most important of the analytical assumptions have been listed in this section.

This analysis has treated only desired signals and thermal system noise. No attempt has been made to treat intermodulation effects or equipment nonlinearities. System nonlinearities, such as limiter effects or the effects of modulation restrictive detection, have been treated.

It is not expected that these channel equations will yield results of absolute accuracy. Rather, ease of handling channel predictions has been obtained with tolerable accuracy through the use of simplifying assumptions. The philosophy has been adopted that the performance of the system analyzed here may be measured in the laboratory. The accuracy of the predictional equations having been determined, a required channel performance margin may be employed for the purposes of predicting for other transmission modes and conditions than those measured in the laboratory.

1.2.2 Method of Presentation

This document is, in a certain sense, tutorial, and in another sense, a working document. Much material, which has been basically derived elsewhere, has been extended or modified and reproduced here. Enough material has been included to make the document almost self-sufficient for the purpose of making performance calculations on the system. The scheme employed in the writing of this document has been to present all basic derivations in appendix form. The main body of the paper was reserved for assembling the individual channel equations from component equations appearing in the appendices. In this manner, the main body of the text is useful for working computations, and the appendices provide the required analytical support.

1.2.3 Assumptions

The most important of the simplifying assumptions which appear throughout the document are tabulated below as an aid to the reader.

    a. Modulation restrictive phase-locked loops operate with no modulation error, except for Doppler effects.

    b. Modulation tracking phase-locked loops operate linearly with respect to phase.

    c. Thresholds for phase-locked devices may be defined on a linear basis after the method of Martin (ref. 7).

    d. All predetection and postdetection filters are ideal with flat amplitude transmission characteristics and square cut-off frequency characteristics.

    e. Input noise to all channels is characterized as being band-limited, white, and Gaussian.

    f. All amplitude limiters are ideal snap-action with ideal pre-limiting and postlimiting filters of equal bandwidth.

    g. All digital waveforms have an ideal square shape.

    h. All output data signal-to-noise ratios are derived for channel demodulators above threshold.

    i. Proper signal design insures no in-channel intermodulation products.

1.3 System Description

A physical description of the system concept, spacecraft and ground equipment, signal design, and system operation of the unified S-band system has been discussed in volume I of this series, NASA TN D-2208. Although the present volume contains block diagrams of the system channels, it is recommended that the reader familiarize himself with volume I prior to reading this volume.

2.0 GROUND-TO-SPACECRAFT CHANNEL ANALYSES

The signal transmitted from the ground to the spacecraft is taken as a sinusoidal carrier phase modulated by a sum of ranging code, up-data subcarrier, and voice subcarrier. This composite signal is demodulated at the spacecraft, and the baseband signal is routed to the premodulation processor for recovery of the subcarriers, and also to the PM modulator for down-link transmission. In this section, we will present the analyses of the transponder carrier tracking channel, and also the voice and up-data channels.

2.1 Carrier Tracking Channel

The performance criterion of the spacecraft carrier-tracking channel is the threshold of the carrier-tracking phase lock-loop. The performance of such a loop has been analyzed in appendix C.4.

The input signal power at the spacecraft is obtained from equation A.3.(7), page A-8, as

$$S_c = \frac{A^2}{2} \cos^2(\Delta\varphi_r) \prod_{j=1}^{L} J_0^2(\Delta\varphi_j) \tag{1}$$

where

A = signal amplitude

$\Delta\phi_r$ = phase deviation of the up-link carrier by the range code

$\Delta\phi_j$ = phase deviation of the up-link carrier by the jth subcarrier

The input noise power, computed in a bandwidth equal to the carrier tracking loop noise bandwidth, B_N, is given by

$$N_c]_{B_N} = |\Phi_{n_i}|^2 2B_N \tag{2}$$

where

$\left|\Phi_{n_i}\right|$ = the magnitude of the flat input noise spectral density.

From equations (1) and (2), we obtain

$$\left.\frac{S_c}{N_c}\right]_{B_N} = \cos^2(\Delta\varphi_r) \prod_{j=1}^{L} J_o^2(\Delta\varphi_i) \left[\frac{S_i}{N_i}\right]_{B_N} \quad (3)$$

where

$$\left.\frac{S_i}{N_i}\right]_{B_N} = \frac{\frac{A^2}{2}}{2\left|\Phi_{n_i}\right| B_N}$$

The threshold value of $\left.\frac{S_c}{N_c}\right]_{B_N}$ may be determined according to the desired specification of the probability of loss of carrier phase lock. The reader may refer to appendix C-4 where $\left.\frac{S_c}{N_c}\right]_{B_N}$ versus the probability of loss of carrier phase lock is treated.

2.2 The Voice Channel

The spacecraft voice channel is shown in figure 2.2-1 along with the up-data channel. Let the narrowest bandwidth prior to the wide-band detector be denoted as B_{LP}. Then the input signal-to-noise ratio computed in B_{LP} is

$$\left.\frac{S_i}{N_i}\right]_{B_{LP}} = \frac{\frac{A^2}{2}}{2\left|\Phi_{n_i}\right| B_{LP}} \quad (1)$$

where

A = signal amplitude

$\left|\Phi_{n_i}\right|$ = the magnitude of the flat input noise spectral density

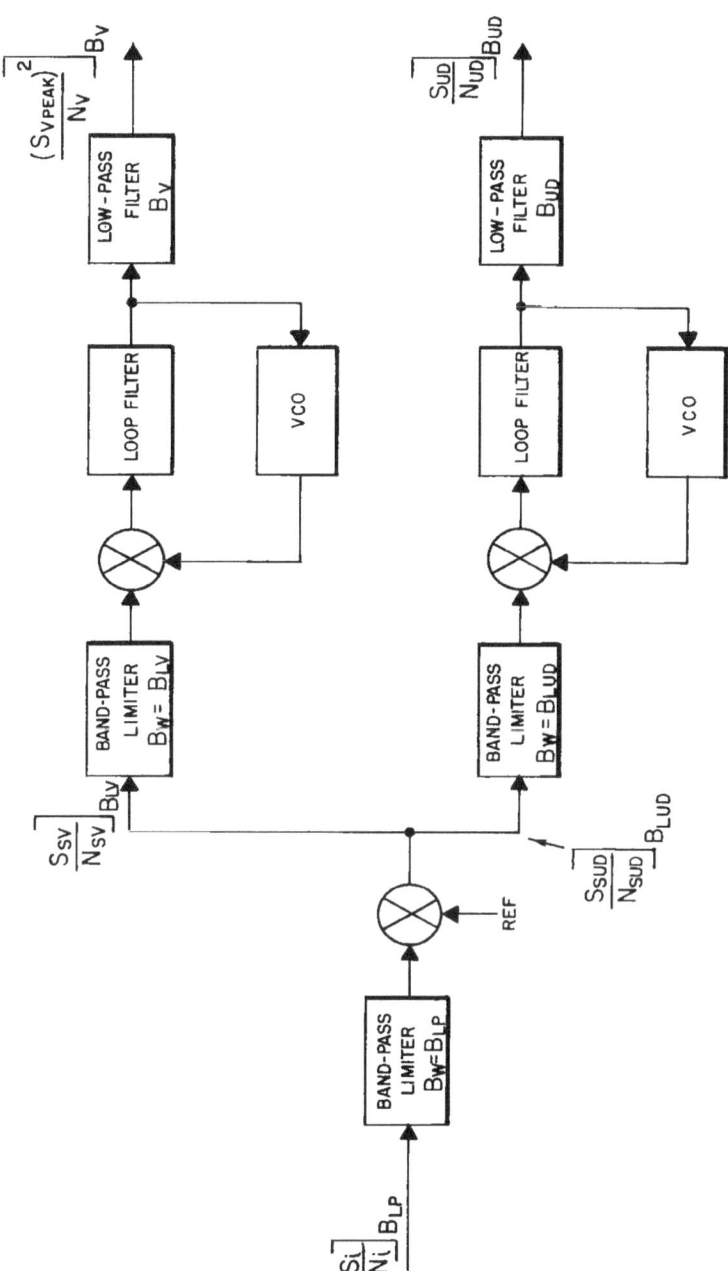

Figure 2.2-1.- Up-link channels

From equation D.1.4.1 (6), page D-7, we obtain the voice subcarrier signal-to-noise ratio computed in B_{LV}, the bandwidth of the band-pass limiter of the voice channel. Thus, from this appendix, and taking into consideration the spacecraft limiter suppression (as treated in section H.3), we obtain

$$\left.\frac{S_{SV}}{N_{SV}}\right]_{B_{LV}} = 2\alpha_L^2 \cos^2\left(\Delta\varphi_r\right) J_1^2\left(\Delta\varphi_V\right) \prod_{\substack{j=1\\j\neq V}}^{L} J_0^2\left(\Delta\varphi_j\right) \left[\frac{B_{LP}}{B_{LV}}\right]\left[\frac{S_i}{N_i}\right]_{B_{LP}} \quad (2)$$

where

α_L = limiter suppression factor

$\Delta\phi_r$ = phase deviation of the carrier by the range code

$\Delta\phi_V$ = phase deviation of the carrier by the voice subcarrier

$\Delta\phi_j$ = phase deviation of the carrier by the jth subcarrier

B_{LV} = voice subcarrier demodulator predetection bandwidth

The quality of the voice channel may be determined by computing the voice information peak-squared signal to mean-squared noise ratio. This ratio was chosen because much work has been performed (reference 23) relating the peak-squared signal to mean-squared noise ratio to intelligibility with clipping depth as a parameter. Thus, from equation E.3.2(6), page E-9, we have

$$\left.\frac{\left(S_{V\,peak}\right)^2}{N_V}\right]_{B_V} = 3\left[\frac{B_{LV}}{B_V}\right]\left[\frac{\Delta f_{V\,peak}}{B_V}\right]^2 \left.\frac{S_{SV}}{N_{SV}}\right]_{B_{LV}} \quad (3)$$

Using now equations (2) and (3), we obtain

$$\left.\frac{\left(S_{V\,peak}\right)^2}{N_V}\right]_{B_V} = 6\alpha_L^2 \left[\frac{B_{LP}}{B_V}\right]\left[\frac{\Delta f_{r\,peak}}{B_V}\right]^2 \cos^2\left(\Delta\varphi_r\right) J_1^2\left(\Delta\varphi_V\right) \prod_{\substack{j=1\\j\neq V}}^{L} J_0^2\left(\Delta\varphi_j\right)\left[\frac{S_i}{N_i}\right]_{B_{LP}}$$

$$(4)$$

Since

$$\left[\frac{B_{LP}}{B_V}\right]\left[\frac{S_i}{N_i}\right]_{B_{LP}} \equiv \left[\frac{S_i}{N_i}\right]_{B_V} \quad (5)$$

equation (4) becomes

$$\left.\frac{(S_{V\;peak})^2}{N_V}\right]_{B_V} = 6\alpha_L^2 \left[\frac{\Delta f_{r\;peak}}{B_V}\right]^2 \cos^2(\Delta\varphi_r) J_1^2(\Delta\varphi_V) \prod_{\substack{j=1 \\ j \neq V}}^{L} J_0^2(\Delta\varphi_j) \left[\frac{S_i}{N_i}\right]_{B_V} \quad (6)$$

where

$\Delta f_{r\;peak}$ = peak frequency deviation of the voice subcarrier by its information

B_V = bandwidth of voice channel postdetection filter

Equation (6) gives the voice signal-to-noise ratio in terms of the range code, the voice channel parameters, and L subcarriers transmitted to the spacecraft.

2.3 The Up-Data Channel

Referring again to figure 2.2-1, the input signal-to-noise ratio (SNR) computed in B_{LP} is

$$\left.\frac{S_i}{N_i}\right|_{B_{LP}} = \frac{\frac{A^2}{2}}{2\left|\Phi_{n_i}\right|B_{LP}} \quad (1)$$

From equation D.1.4.1 (6), page D-7, we obtain the up-data subcarrier SNR computed in B_{LUD}, the bandwidth of the band-pass limiter of the up-data channel.

Thus

$$\left.\frac{S_{SUD}}{N_{SUD}}\right]_{B_{LUD}} = 2\alpha_L^2 \cos^2(\Delta\varphi_r) J_1^2(\Delta\varphi_{UD}) \prod_{\substack{j=1 \\ j \neq UD}}^{L} J_0^2(\Delta\varphi_j) \left[\frac{B_{LP}}{B_{LUD}}\right]\left[\frac{S_i}{N_i}\right]_{B_{LP}} \quad (2)$$

As in the case of the voice channel, the peak-squared signal to mean-squared noise ratio will be used to determine the up-data channel quality. Thus, from equation E.3.2 (6), page E-11, we have

$$\left[\frac{S_{UD}}{N_{UD}}\right]_{B_{UD}} = 3K_p^2 \left[\frac{B_{LUD}}{B_{UD}}\right]\left[\frac{\Delta f_{UD\ peak}}{B_{UD}}\right]^2 \left[\frac{S_{SUD}}{N_{SUD}}\right]_{B_{LUD}} \quad (3)$$

where

$K_p = \dfrac{\text{rms}}{\text{peak}}$ of the up-data signal.

Using now equations (2), (3), and the fact that

$$\left[\frac{B_{LP}}{B_{UD}}\right]\left[\frac{S_i}{N_i}\right]_{B_{LP}} \equiv \left[\frac{S_i}{N_i}\right]_{B_{UD}}$$

we obtain

$$\left[\frac{S_{UD}}{N_{UD}}\right]_{B_{UD}} = 6\,\alpha_L^2 K_p^2 \left[\frac{\Delta f_{UD\ peak}}{B_{UD}}\right]^2 \cos^2(\Delta\varphi_r)\, J_1^2(\Delta\varphi_{UD}) \prod_{\substack{j=1 \\ j \neq UD}}^{L} J_0^2(\Delta\varphi_j) \left[\frac{S_i}{N_i}\right]_{B_{UD}}$$

(4)

where

α_L = limiter supression factor

$\Delta f_{UD\ peak}$ = peak frequency deviation of the up-data subcarrier by its information

B_{UD} = bandwidth of the up-data channel postdetection filter

$\Delta\varphi_r$ = phase deviation of the carrier by the range code

$\Delta\varphi_{UD}$ = phase deviation of the carrier by the up-data subcarrier

$\Delta\varphi_j$ = phase deviation of the carrier by the jth subcarrier

since the up-data receiver is a subsystem separate from the spacecraft S-band subsystem, it is assumed that the output data quality of the up-data receiver may be uniquely related to the output signal-to-noise ratio of the up-data subcarrier demodulator, $\left.\frac{S_{UD}}{N_{UD}}\right]_{B_D}$. Given a specification for $\left.\frac{S_{UD}}{N_{UD}}\right]_{B_{UD}}$, equation (4) may be used to infer the channel quality.

3.0 SPACECRAFT-TO-GROUND PHASE MODULATED CHANNEL ANALYSES

3.1 Carrier Tracking Channel

There are two performance criteria for the carrier tracking channel. One criterion is the input signal-to-noise ratio at which the channel thresholds. The other criterion is the input signal-to-noise ratio at which the VCO phase jitter is acceptable for Doppler tracking.

In section F.3, it was shown that the ground carrier tracking loop may be treated for threshold as in section C.4, given a knowledge of the loop's equivalent threshold noise bandwidth B_N.

The ground-received signal power in the receiver closed loop noise bandwidth is obtained from equation F.5.2 (7), page F-36, as

$$S_c = \frac{A_g^2}{2} e^{-\sigma_{\varphi s}^2} \cos^2(\Delta\varphi_r \text{eff}) \prod_{j=1}^{L} J_o^2(\Delta\varphi_j \text{eff}) \prod_{i=1}^{K} J_o^2(\Delta\varphi_i) \quad (1)$$

where

$e^{-\sigma_{\varphi s}^2}$ = signal suppression factor due to phase modulated noise in the spacecraft turnaround channel

A_g = peak value of the received sinusoidal carrier

$$\Delta\varphi_r \text{eff} = \Delta\varphi_m \alpha_{LS} \sin(\Delta\varphi_r) \prod_{j=1}^{L} J_o(\Delta\varphi_j) \quad (2)$$

and

$$\Delta\varphi_j\text{eff} = 2\Delta\varphi_m \alpha_{LS} \cos(\Delta\varphi_r) J_1(\Delta\varphi_j) \prod_{\substack{h=1 \\ h \neq j}}^{L} J_0(\Delta\varphi_h) \tag{3}$$

The reader should refer to section F.5 for the discussion of the spacecraft turnaround channel and the derivation of equations (2) and (3). Other terms are defined as:

$\Delta\varphi_m$ = spacecraft turnaround channel phase gain, neglecting limiter suppression

$\Delta\varphi_r\text{eff}$ = effective phase deviation of the carrier by the turnaround range code

$\Delta\varphi_j$, $\Delta\varphi_h$ = subcarrier phase deviation on the up-link carrier

$\Delta\varphi_i$ = subcarrier phase deviation on the down-link carrier

α_{LS} = spacecraft limiter signal suppression factor

$\Delta\varphi_j\text{eff}$ = effective phase deviation of the carrier by the turnaround subcarriers

The ground receiver noise spectral density is attributed to the normal system noise spectral density $\left|\Phi_{n_i}\right|$ plus the phase noise transmitted from the spacecraft during the turnaround process. The total receiver noise spectral density is defined here as $\left|\Phi_{n_T}\right|$ and it is treated in detail in section F.5.1. Thus, the noise in the ground receiver closed loop noise bandwidth B_N is

$$N_c = 2 \left|\Phi_{n_T}\right| B_N \tag{4}$$

The signal-to-noise ratio computed in the closed loop noise bandwidth may be obtained from (1) and (4). Thus, we have

$$\left.\frac{S_c}{N_c}\right]_{B_N} = \frac{\dfrac{A_g^2}{2} e^{-\sigma_{\varphi s}^2} \cos^2(\Delta\varphi_r\text{eff}) \prod_{j=1}^{L} J_0^2(\Delta\varphi_j\text{eff}) \prod_{i=1}^{K} J_0^2(\Delta\varphi_i)}{2 \left|\Phi_{n_T}\right| B_N} \tag{5}$$

Since

$$\left.\frac{S_{ig}}{N_{ig}}\right]^{\Phi_{n_T}}_{B_N} = \frac{e^{-\sigma^2_{\varphi s}} A_g^2}{2\left|\Phi_{n_T}\right| B_N}$$

equation (5) may be simplified. Thus

$$\left.\frac{S_c}{N_c}\right]_{B_N} = \cos^2(\Delta\varphi_r \text{eff}) \prod_{j=1}^{L} J_0^2(\Delta\varphi_j \text{eff}) \prod_{i=1}^{K} J_0^2(\Delta\varphi_i) \left[\frac{S_{ig}}{N_{ig}}\right]^{\Phi_{n_T}}_{B_N} \quad (6)$$

The threshold value of $\left.\frac{S_c}{N_c}\right]_{B_N}$ may be specified as in section C.4.

3.2 Angle Tracking Channel

The closed loop noise bandwidth of the angle channel is considerably smaller than that of the carrier tracking channel. Since the angle channel depends on the carrier channel VCO for phase reference, the angle channel performance is directly tied to the carrier channel performance. In particular, the angle channel does not perform when the carrier channel thresholds. Therefore, equation 3.1 (5) may also be used to define angle channel threshold.

3.3 Ranging Channel

The basic model required to analyze the performance of the ranging channel is shown in figure 3.3-1. Since the ranging channel includes the spacecraft transponder as well as the ground range clock receiver, the reader is urged to read sections F.1 and F.5.

Various terms are defined below:

S_{is} = input signal power to the spacecraft receiver

$\Phi_{n_{is}}(\omega)$ = spacecraft input noise spectral density

$\left.\frac{S_{is}}{N_{is}}\right]_{B_{LS}}$ = spacecraft signal-to-noise ratio at limiter input, computed in limiter bandwidth

α_{LS} = spacecraft band-pass limiter signal suppression factor

B_{LS} = spacecraft band-pass limiter bandwidth

V_{LS} = spacecraft band-pass limiter voltage limiting level

K'_{φ} = spacecraft wideband detector gain constant

$\Delta\varphi_m$ = spacecraft turnaround channel phase gain constant

S_{os} = spacecraft output carrier power

S_{ig} = input signal power to the ground receiver

$\Phi_{n_{ig}}(\omega)$ = ground thermal input noise spectral density

$\Phi_{n_T}(\omega)$ = ground total "equivalent" input noise spectral density, including turned around noise

$\left.\dfrac{S_{ig}}{N_{ig}}\right]^{\Phi_{n_T}}$ = ground signal-to-noise ratio at range clock receiver input, computed using total equivalent noise spectral density

$\left.\dfrac{S_c}{N_c}\right]_{B_{LR}}$ = clock signal-to-noise ratio at limiter input, computed in limiter bandwidth

B_{LR} = range clock receiver limiter bandwidth

B_N = range clock loop noise bandwidth

There are two performance criteria for the ranging channel. One concerns the signal-to-noise ratio (SNR), required at the input of the range clock receiver to insure that the clock loop (see figure F.1-1) is above threshold. The second criterion concerns the input SNR required for a given range code acquisition time.

3.3.1 Clock Loop Threshold

Given a specification for the probability of loss of lock, the threshold properties of the range clock loop are implied by the clock SNR at the limiter input, $\left.\dfrac{S_c}{N_c}\right]_{B_N}$, computed in B_N, the clock loop noise bandwidth. Combining equation F.1.4 (1), page F-12, and F.5.2 (4), page F-36, we obtain

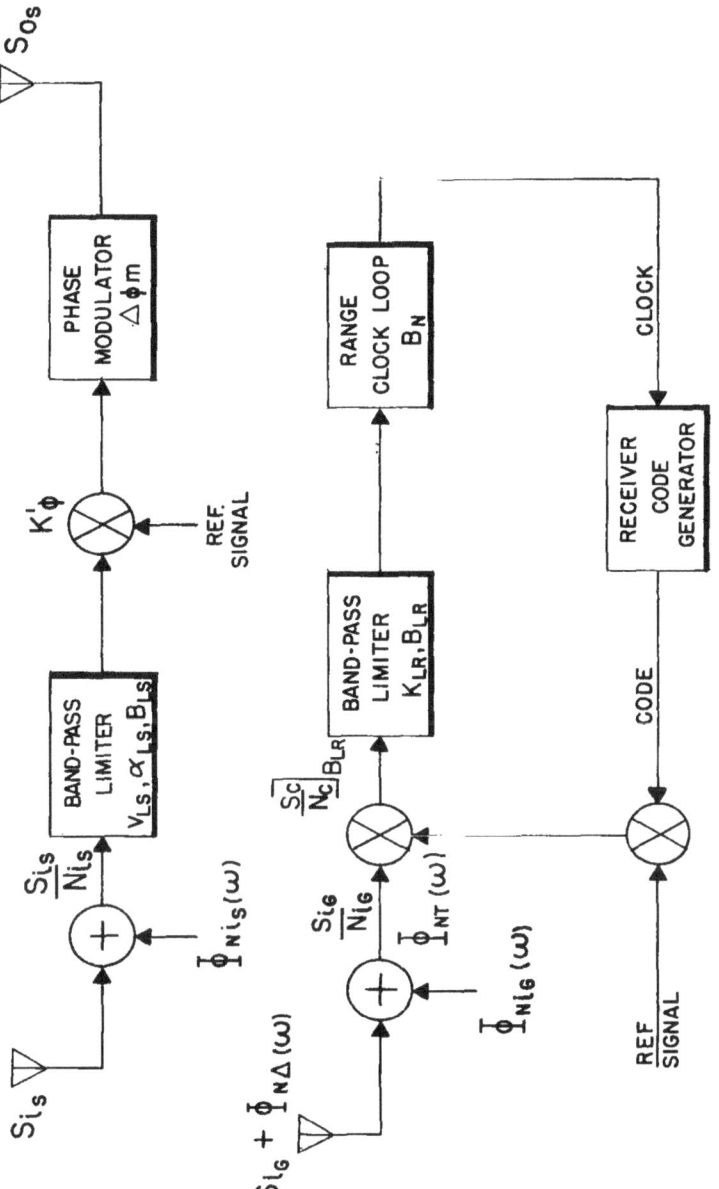

Figure 3.3-1.- The ranging channel

$$\left.\frac{S_c}{N_c}\right]_{B_N} = L_D L_K \sin^2(\Delta\varphi_j \text{eff}) \prod_{i=1}^{L} J_o^2(\Delta\varphi_j \text{eff}) \prod_{i=1}^{K} J_o^2(\Delta\varphi_i) \left[\frac{S_{ig}}{N_{ig}}\right]_{B_N}^{\Phi_{n_T}} \tag{1}$$

where

$\Delta\varphi_r \text{eff}$ = The equivalent turnaround phase deviation of the range code on the down carrier as given by equation F.5.2 (5), page F-36.

$\Delta\varphi_j \text{eff}$ = The equivalent turned-around phase deviation of the up-subcarriers on the down carrier as given by equation F.5.2 (6), page F-36.

L_D = a signal detection loss as defined in section F.1.3, page F-12.

L_K A correlation loss as defined in section F.1.3, page F-11.

The input SNR computed in B_N is given as

$$\left.\frac{S_{ig}}{N_{ig}}\right]_{B_N}^{\Phi_{n_T}} = \frac{\frac{A_g^2}{2} e^{-\sigma_{\varphi s}^2}}{\left|\Phi_{n_T}\right|^2 B_N} \tag{2}$$

where

A_g = the peak value of the sinusoidal carrier received at the ground

$\sigma_{\varphi s}^2$ = the mean-squared value of the turned-around phase noise as given by equation F.5.1 (5), page F-34.

The value of $\left|\Phi_{n_T}\right|$ may be obtained from equation F.5.1 (12), page F-33.

Using equation (1) and (2), the range clock loop may be treated for threshold as in section C.4.

3.3.2 Range Code Acquisition Time

From figure F.1.5-1, page F-17, the acquisition time for the pseudo-random ranging code may be directly related to the ratio of effective output signal-to-noise spectral density $\dfrac{S_o}{|\Phi_o|}$, using equation F.1.5 (14), page F-16.

Combining equations F.1.5 (10), page F-15, and F.5 (4), page F-30, we obtain

$$\frac{S_o}{|\Phi_o|} = \frac{1}{6.12\pi} \sin^2(\Delta\varphi_r \text{eff}) \prod_{j=1}^{L} J_o^2(\Delta\varphi_j \text{eff}) \prod_{i=1}^{K} J_o^2(\Delta\varphi_i) \frac{S_{ig}}{|\Phi_{n_T}|} \qquad (1)$$

where

$$S_{ig} = \frac{A_g^2}{2} e^{-\sigma_{\varphi s}^2} \qquad (2)$$

and the quantities A_g, $\sigma_{\varphi s}$, $\Delta\varphi_r \text{eff}$, $|\Phi_{n_T}|$, and $\Delta\varphi_j \text{eff}$ are the same as in section 3.3.1.

3.4 The PCM Telemetry Channel

The PCM telemetry channel is shown in figure 3.4-1. The figure includes only those components of the channel necessary for the analysis to follow.

Let us again define the channel input signal-to-noise ratio as

$$\left.\frac{S_{ig}}{N_{ig}}\right]_{B_{LP}}^{\Phi_{n_T}} = \frac{e^{-\sigma_{\varphi s}^2} \dfrac{A_g^2}{2}}{2|\Phi_{n_T}| B_{LP}} \qquad (1)$$

where as before

A_g = amplitude of the sinusoidal carrier received at the ground

B_{LP} = band-pass filter bandwidth

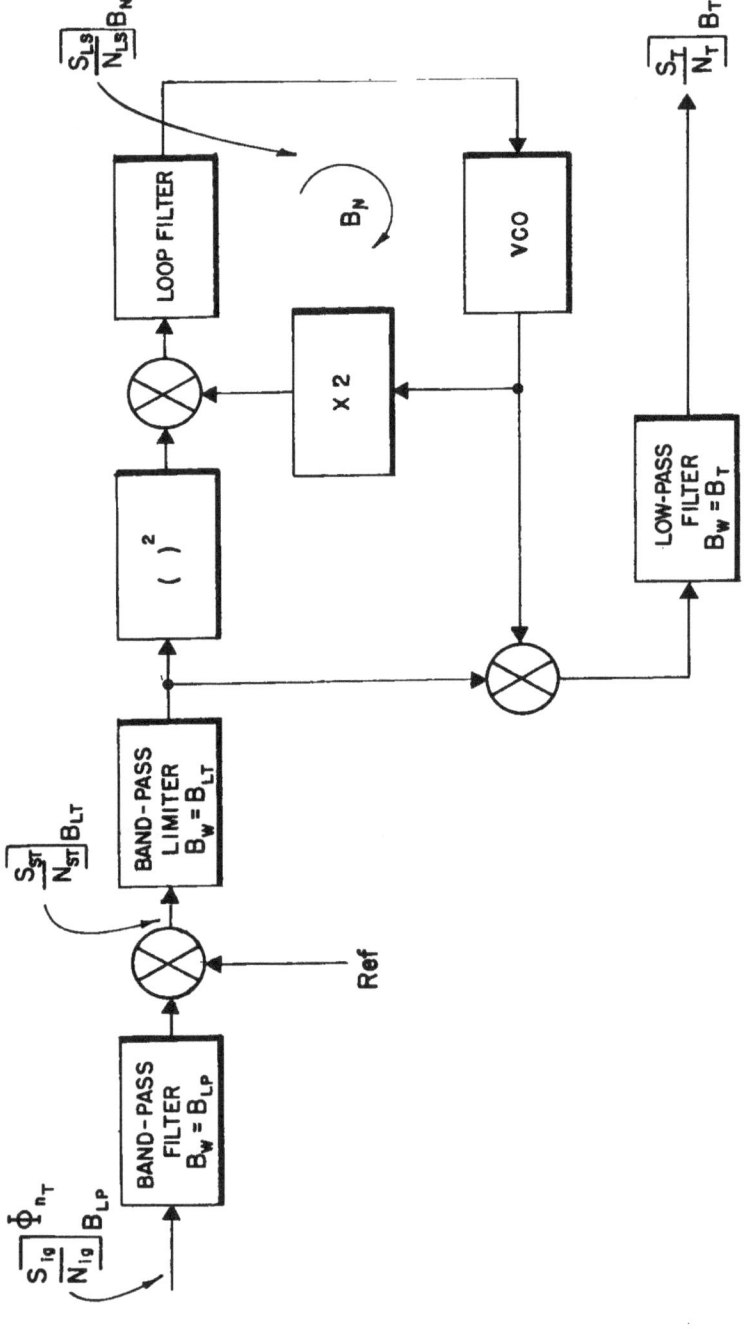

Figure 3.4-1.- PCM telemetry channel

$\Phi_{n_T}(f)$ = total effective "input" noise spectral density

At this point the reader should note carefully that

$$\Phi_{n_T}(f) \equiv \Phi_{n_i}(f)$$

only when the transponder ranging channel is closed. When this channel is open, $\Phi_{n_T}(f)$ includes the transponder turned-around noise as well as the "normal" ground system input noise. The term $\Phi_{n_T}(f)$ is given in appendix F.6.1.

The recovery of a subcarrier using a band-pass filter has been treated in appendix D. Thus, from equation D.1.4.1 (6), page D-7, and appendix F, equation F.5.2 (4), we obtain

$$\left.\frac{S_{ST}}{N_{ST}}\right]_{B_{LT}} = 2\cos^2(\Delta\varphi_r\text{eff})\, J_1^2(\Delta\varphi_T) \prod_{\substack{i=1 \\ i \neq T}}^{K} J_0^2(\Delta\varphi_i) \prod_{j=1}^{L} J_0^2(\Delta\varphi_j\text{eff}) \left[\frac{B_{LP}}{B_{LT}}\right] \left[\frac{S_{ig}}{N_{ig}}\right]_{B_{LP}}^{\Phi_{n_T}} \tag{2}$$

where $\Delta\varphi_r\text{eff}$ and $\Delta\varphi_j\text{eff}$ are as defined in section 3.1 and

$e^{-\sigma_{\varphi_S}^2}$ = signal suppression factor due to phase modulation of the down-link carrier by transponder turned-around noise.

Now

$$\left.\frac{S_T}{N_T}\right]_{B_T} = \left[\frac{B_{LT}}{B_T}\right] \left[\frac{S_{ST}}{N_{ST}}\right]_{B_{LT}} \tag{3}$$

Combining now equations (2) and (3), we obtain

$$\left.\frac{S_T}{N_T}\right]_{B_T} = 2\cos^2(\Delta\varphi_r\text{eff})\, J_1^2(\Delta\varphi_T) \prod_{\substack{i=1 \\ i \neq T}}^{K} J_0^2(\Delta\varphi_i) \prod_{j=1}^{L} J_0^2(\Delta\varphi_j\text{eff}) \left[\frac{B_{LP}}{B_T}\right] \left[\frac{S_{ig}}{N_{ig}}\right]_{B_{LP}}^{\Phi_{n_T}} \tag{4}$$

Considering now that

$$\left[\frac{B_{LP}}{B_T}\right]\left[\frac{S_{ig}}{N_{ig}}\right]_{B_{LP}} = \left[\frac{S_{ig}}{N_{ig}}\right]_{B_T}$$

equation (4) becomes

$$\left.\frac{S_T}{N_T}\right]_{B_T} = 2\cos^2(\Delta\varphi_r \text{eff})J_1^2(\Delta\varphi_T)\prod_{\substack{i=1\\i\neq T}}^{K} J_0^2(\Delta\varphi_i)\prod_{j=1}^{L} J_0^2(\Delta\varphi_j \text{eff})\left[\frac{S_{ig}}{N_{ig}}\right]_{B_T}^{\Phi_{n_T}}$$

(5)

The telemetry demodulator in this report has been considered as a specialized detector. As such, it has been treated in section F.3. The reader may refer to this section for discussion of the demodulator threshold.

3.5 The Voice Channel

The voice information is transmitted from the spacecraft on a subcarrier which is also used for transmission of biomedical data. Thus, the ground subcarrier demodulator is common to both voice and biomedical data channels. The voice channel is shown in figure 3.5-1. The channel input signal-to-noise ratio is defined as

$$\left.\frac{S_{ig}}{N_{ig}}\right|_{B_{LP}}^{\Phi_{n_T}} = \frac{e^{-\sigma_{\varphi s}^2}\frac{A_g^2}{2}}{2\left|\Phi_{n_T}\right|B_{LP}} \quad (1)$$

where

A_g = amplitude of the signal

B_{LP} = band-pass filter bandwidth

$\left|\Phi_{n_T}\right|$ = input noise spectral density

As in the case of PCM telemetry $\left|\Phi_{n_T}\right| \equiv \left|\Phi_{n_i}\right|$ only when the transponder ranging channel is closed. Otherwise, it is defined in section F.6.

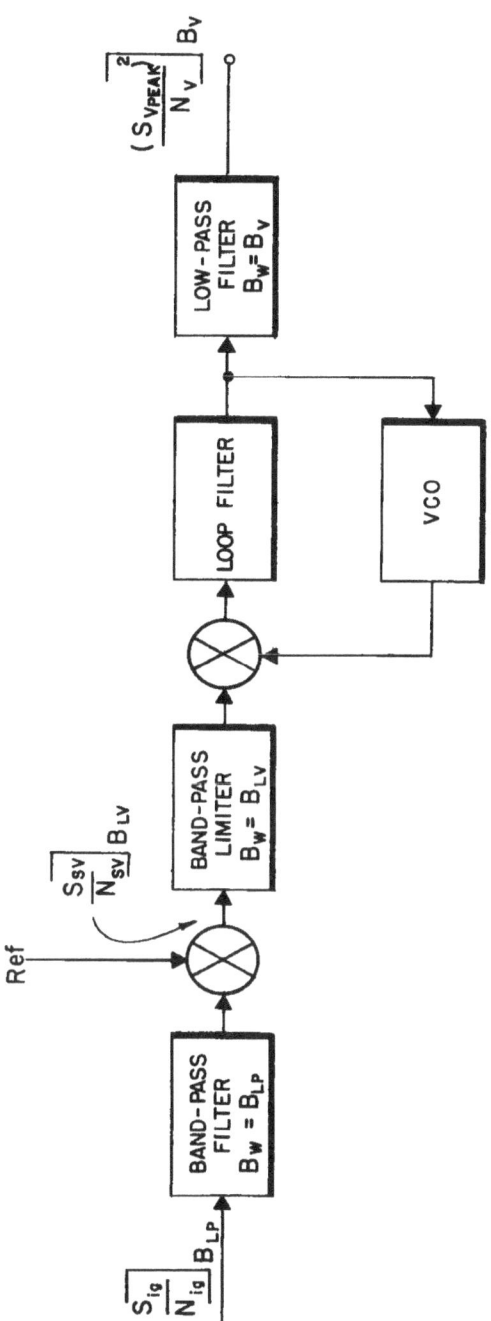

Figure 3.5-1.- The voice channel

The subcarrier signal-to-noise ratio may be obtained from appendix D.1.4.1, equation (6). Thus

$$\left.\frac{S_{SV}}{N_{SV}}\right]_{B_{LV}} = 2\cos^2(\Delta\varphi_r\text{ eff})\,J_1^2(\Delta\varphi_V)\prod_{\substack{i=1\\i\neq V}}^{K}J_0^2(\Delta\varphi_i)\prod_{j=1}^{L}J_0^2(\Delta\varphi_j\text{eff})\left[\frac{B_{LP}}{B_{LV}}\right]\left[\frac{S_{ig}}{N_{ig}}\right]_{B_{LV}}$$

(2)

As shown in figure 3.5-1, the voice information is recovered with a low-pass filter at the output of the modulation tracking loop. The voice channel quality may be determined by computing the peak-squared signal to mean-squared noise ratio at the output of the voice channel low-pass filter. This ratio was chosen by the authors, so that it may be used directly to evaluate voice intelligibility.

Now, from section E.3.2 (6), page E-11, we have

$$\left.\frac{(S_{V\text{ peak}})^2}{N_V}\right]_{B_{LV}} = 3\left[\frac{\Delta f_{V\text{ peak}}}{B_V}\right]^2\left[\frac{B_{LV}}{B_V}\right]\left[\frac{S_{SV}}{N_{SV}}\right]_{B_{LV}}$$

(3)

Using now equations (2) and (3), we obtain

$$\left.\frac{(S_{V\text{ peak}})^2}{N_V}\right]_{B_V} = 6\left[\frac{\Delta f_{r\text{ peak}}}{B_V}\right]^2\cos^2(\Delta\varphi_r\text{ eff})\,J_1^2(\Delta\varphi_V)\prod_{\substack{i=1\\i\neq V}}^{K}$$

$$J_0^2(\Delta\varphi_i)\prod_{j=1}^{L}J_0^2(\Delta\varphi_j\text{eff})\left[\frac{B_{LP}}{B_V}\right]\left[\frac{S_{ig}}{N_{ig}}\right]_{B_{LP}}^{\Phi_{n_T}}$$

(4)

Now since

$$\left[\frac{B_{LP}}{B_V}\right]\left[\frac{S_{ig}}{N_{ig}}\right]_{B_{LP}} \equiv \left[\frac{S_{ig}}{N_{ig}}\right]_{B_V}$$

(5)

$$\left.\frac{(S_{V\;peak})^2}{N_V}\right]_{B_V} = 6\left[\frac{\Delta f_{V\;peak}}{B_V}\right]^2 \cos^2(\Delta\varphi_r\text{eff})\, J_1^{\;2}(\Delta\varphi_V) \prod_{\substack{i=1 \\ i \neq V}}^{K}$$

$$J_o^{\;2}(\Delta\varphi_i) \prod_{j=1}^{L} J_o^{\;2}(\Delta\varphi_j\text{eff}) \left[\frac{S_{ig}}{N_{ig}}\right]_{B_V}^{\Phi_{n_T}}$$

(6)

where

$\Delta f_{V\;peak}$ = peak frequency deviation of the voice subcarrier by its information

B_V = postdetection filter bandwidth

$\Delta\varphi_V$ = phase deviation of the carrier by the voice subcarrier

$\Delta\varphi_r\text{eff}$ = phase deviation of the carrier by the range code as defined in appendix E.6

$\Delta\varphi_j\text{eff}$ = phase deviation of the carrier by the jth subcarrier turned around in the spacecraft as defined in appendix E.6

$\Delta\varphi_i$ = phase deviation of the carrier by the ith subcarrier originating in the spacecraft

The voice channel threshold may be treated as in appendix C.

3.6 Biomedical Data Channels

The seven biomedical data channels are identical. Therefore, only one of them will be analyzed here. One of these channels is shown in figure 3.6-1.

Examination of figure 3.5-1, section 3.5, reveals that equations (1) and (2) of the voice channel are the same for the biomedical data channels. We may then proceed with the signal-to-noise ratio of one biomedical data subcarrier at the output of the voice and biomedical data subcarrier modulation tracking loop. Since the band-pass filter is

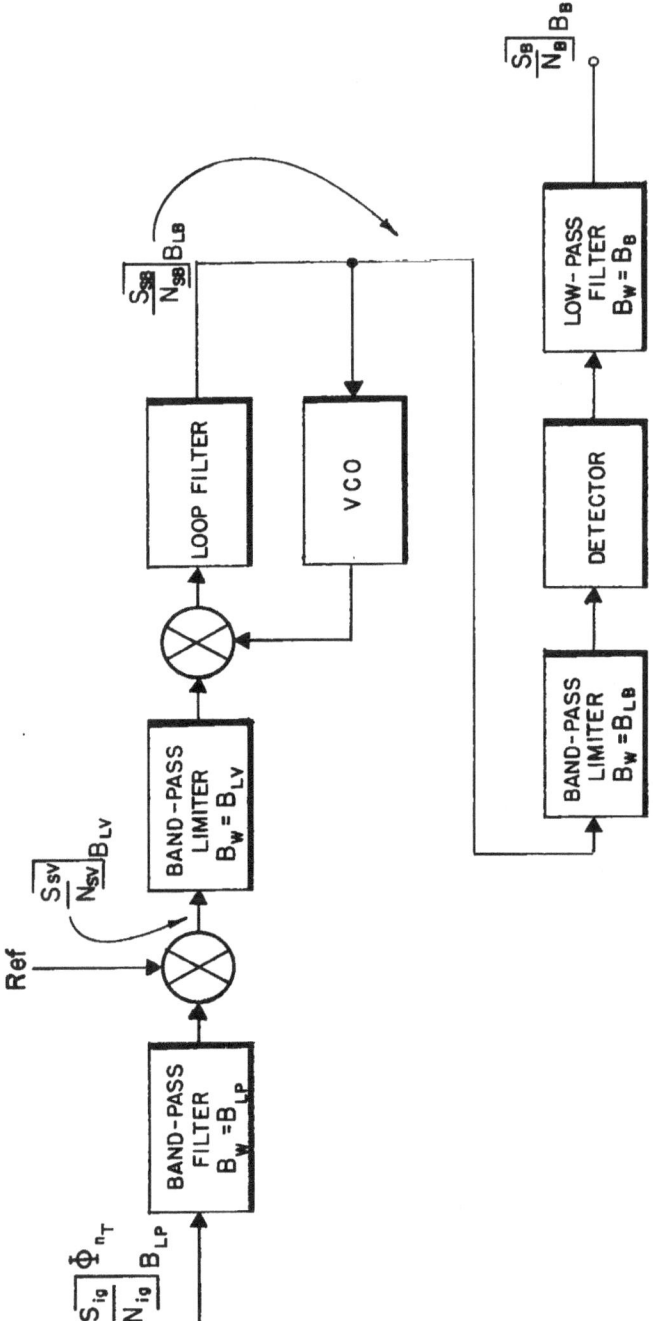

Figure 3.6-1.- The biomedical data channel

used for the recovery of the subcarrier, we may refer to appendix E, section E.3.1, equation (8), page E-10. Thus,

$$\left[\frac{S_{SB}}{N_{SB}}\right]_{B_{LB}} \cong \frac{1}{2}\left[\frac{\Delta f_b}{f_b}\right]^2 \left[\frac{B_{LV}}{B_{LB}}\right]\left[\frac{S_{SV}}{N_{SV}}\right]_{B_{LV}} \qquad (1)$$

where

Δf_b = frequency deviation of the voice and biomedical data subcarrier due to the biomedical data subcarrier in question

f_b = frequency of the biomedical data subcarrier in question

The detector shown in biomedical data channel may be either a modulation tracking loop or an FM discriminator. Excluding threshold considerations, the analysis of this section holds for either type of detector.

The recovery of data from a subcarrier using a modulation tracking loop and a low-pass filter is covered in detail in appendix E. Thus, from section E.3.2, equation (8), we find that

$$\left[\frac{S_B}{N_B}\right]_{B_B} = 3K_p^2 \left[\frac{\Delta f_{b\ peak}}{B_B}\right]^2 \left[\frac{S_{SB}}{N_{SB}}\right]_{B_{LB}} \qquad (2)$$

Using now equations (1) and (2) of section 3.6 and (1) and (2) of this section, we obtain

$$\left[\frac{S_B}{N_B}\right]_{B_B} = 3K_p^2 \left[\frac{(\Delta f_{b\ peak})}{B_B}\right]^2 \left[\frac{\Delta f_b}{f_b}\right]^2 \cos^2(\Delta\varphi_r \text{eff}) J_1^2(\Delta\varphi_V) \prod_{\substack{i=1 \\ i \neq V}}^{K}$$

$$J_o^2(\Delta\varphi_i) \prod_{j=1}^{L} J_o^2(\Delta\varphi_j \text{eff}) \left[\frac{B_{LP}}{B_B}\right]\left[\frac{S_{ig}}{N_{ig}}\right]_{B_{LP}}^{\Phi_{n_T}}$$

(3)

where

K_p = factor relating the peak to rms value of the baseband modulation of the biomedical data subcarrier in question

$\Delta f_{b\ peak}$ = the peak frequency deviation of the biomedical data subcarrier by its modulation

B_B = bandwidth of biomedical data postdetection filter

The rest of the terms of equation (3) are as defined in section 3.6. Now since

$$\left[\frac{B_{LP}}{B_B}\right]\left[\frac{S_{ig}}{N_{ig}}\right]_{B_{LP}} \equiv \left[\frac{S_{ig}}{N_{ig}}\right]_{B_B}$$

equation (3) becomes

$$\left[\frac{S_B}{N_B}\right]_{B_B} = 3K_p^2 \left[\frac{\Delta f_{b\ peak}}{B_B}\right]^2 \left[\frac{\Delta f_b}{f_b}\right]^2 \cos^2(\Delta\varphi_r\,\text{eff})\, J_1^2(\Delta\varphi_V) \prod_{\substack{i=1 \\ i \neq V}}^{K}$$

$$J_0^2(\Delta\varphi_i) \prod_{j=1}^{L} J_0^2(\Delta\varphi_j\,\text{eff}) \left[\frac{S_{ig}}{N_{ig}}\right]_{B_B}^{\Phi_{n_T}} \quad (4)$$

The threshold of the biomedical data channels is essentially that of the voice demodulator since voice and the biomedical data subcarriers are frequency multiplexed. The voice demodulator threshold treatment may be found in appendix D.

3.7 The Emergency Voice Channel

The emergency voice channel is shown in figure 3.7-1. The reader is reminded at this point that the emergency voice signal does not contain the "turn-around" noise. This is because the voice signal is modulated directly on the VCO output while the turn-around ranging channel is inactive during this mode of transmission.

The voice channel input signal-to-noise ratio may be defined as:

$$\left[\frac{S_{ig}}{N_{ig}}\right]_{B_{LP}} = \frac{\frac{A_g^2}{2}}{2\left|\Phi_{n_i}\right| B_{LP}} \quad (1)$$

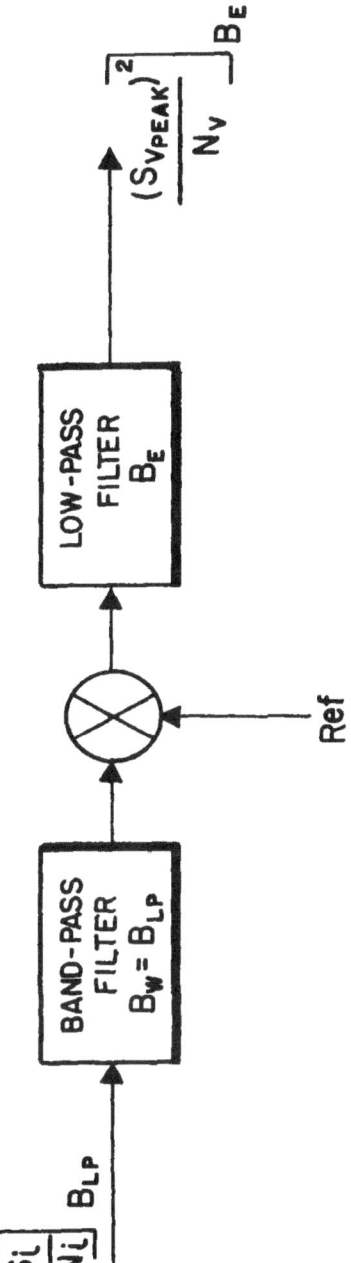

Figure 3.7-1.- The emergency voice channel

where

A_g = amplitude of input signal

B_{LP} = bandwidth of band-pass limiter

$|\Phi_{n_i}|$ = input noise spectral density

Again the peak-squared signal to mean-squared noise will be used for evaluation of the channel quality. Thus, from section D.1.4.2 (5), page D-8

$$\left[\frac{(S_{V\ peak})^2}{N_V}\right]_{B_E} = \sin^2(\Delta\varphi_{V\ peak})\left[\frac{S_{ig}}{N_{ig}}\right]_{B_{LP}} \qquad (2)$$

where

$\Delta\varphi_{V\ peak}$ = peak phase deviation of the carrier by the voice information

Using now equation (2), and taking into consideration the bandwidth ratio, we obtain

$$\left[\frac{(S_{V\ peak})^2}{N_V}\right]_{B_E} = \sin^2(\Delta\varphi_{V\ peak})\left[\frac{B_{LP}}{B_E}\right]\left[\frac{S_{ig}}{N_{ig}}\right]_{B_{LP}} \qquad (3)$$

where

B_E = bandwidth of low-pass filter used to recover the voice information

Now since

$$\left[\frac{B_{LP}}{B_E}\right]\left[\frac{S_{ig}}{N_{ig}}\right]_{B_{LP}} \equiv \left[\frac{S_{ig}}{N_{ig}}\right]_{B_E}$$

equation (3) becomes

$$\left[\frac{(S_{V\ peak})^2}{N_V}\right]_{B_E} = \sin^2(\Delta\varphi_{V\ peak})\left[\frac{S_{ig}}{N_{ig}}\right]_{B_E} \qquad (4)$$

The threshold of the emergency voice demodulator is essentially that of the carrier tracking loop. The carrier tracking loop threshold has been discussed in section 3.1.

3.8 The Emergency Key Channel

The emergency key channel is shown in figure 3.8-1. As the figure indicates, a beat-frequency oscillator is used for demodulation. The reader is reminded that no turn-around noise exists in the channel since the transponder ranging channel is not active during transmission of emergency key.

Again, the channel input signal-to-noise ratio is defined as

$$\left[\frac{S_{ig}}{N_{ig}}\right]_{B_{LP}} = \frac{\frac{A_g^2}{2}}{2\left|\Phi_{n_i}\right|B_{LP}} \tag{1}$$

When present, the emergency key signal appears as a subcarrier in the incoming signal. Thus, from equation D.1.4.1 (5), page D-7, we find that the signal-to-noise ratio of a subcarrier, recovered by a band-pass filter, at the output of a modulation restrictive loop is

$$\left[\frac{S_{SK}}{N_{SK}}\right]_{B_{SK}} = 2J_1^2(\Delta\varphi_K)\left[\frac{B_{LP}}{B_{SK}}\right]\left[\frac{S_{ig}}{N_{ig}}\right]_{B_{LP}} \tag{2}$$

where

$\Delta\varphi_K$ = phase deviation of the carrier by the key subcarrier

However, since

$$\left[\frac{S_K}{N_K}\right]_{B_K} = \left[\frac{B_{SK}}{B_K}\right]\left[\frac{S_{SK}}{N_{SK}}\right]_{B_{SK}} \tag{3}$$

Equation (2) becomes

$$\left[\frac{S_K}{N_K}\right]_{B_K} = 2J_1^2(\Delta\varphi_K)\left[\frac{B_{LP}}{B_K}\right]\left[\frac{S_{ig}}{N_{ig}}\right]_{B_{LP}} \tag{4}$$

Now since

$$\left[\frac{B_{LP}}{B_K}\right]\left[\frac{S_{ig}}{N_{ig}}\right]_{B_{LP}} \equiv \left[\frac{S_{ig}}{N_{ig}}\right]_{B_K}$$

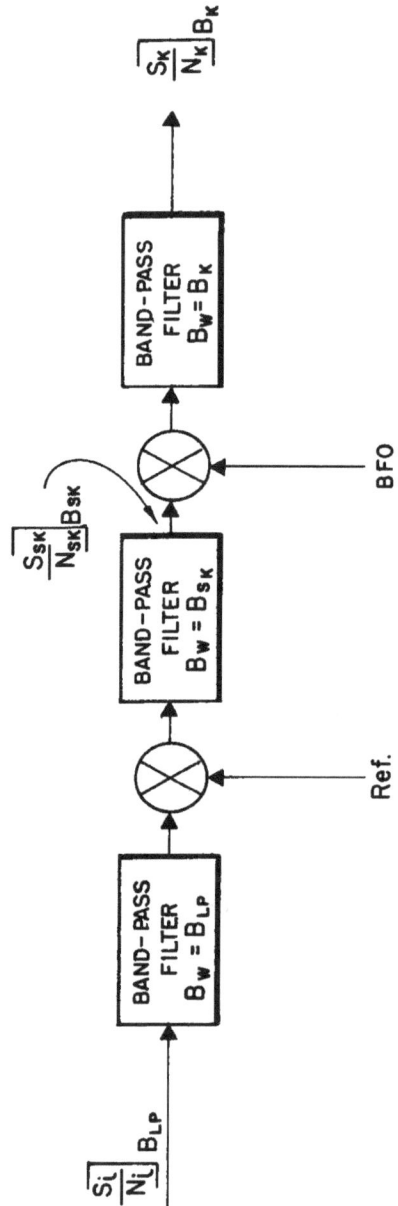

Figure 3.8-1.- The emergency key channel

equation (4) becomes

$$\left.\frac{S_K}{N_K}\right]_{B_K} = 2J_1^2(\Delta\varphi_K)\left[\frac{S_{ig}}{N_{ig}}\right]_{B_K} \qquad (5)$$

where

B_K = key postdetection filter bandwidth

The emergency key channel threshold is essentially the carrier loop threshold. Thus, the reader may refer to section 3.1 for threshold relations of the key channel.

4.0 SPACECRAFT-TO-GROUND FREQUENCY MODULATED CHANNEL ANALYSES

The spacecraft frequency modulated carrier is assumed to contain television at baseband and two subcarriers. One subcarrier is identified as the PCM telemetry subcarrier, and may contain either high or low bit rate real-time telemetry or apparent high bit rate recorded telemetry. The other subcarrier is identified as the voice subcarrier, and may contain either real-time clipped voice plus biomedical data, or recorded voice.

The following sections will treat separately the output data signal-to-noise ratios for television, voice, PCM telemetry, and biomedical data. The reader should note that the mathematical relationships will involve K subcarriers, rather than two, in order that these relationships remain general.

4.1 Carrier Demodulation Channel

The performance criterion for the carrier demodulator, which is a modulation tracking phase-locked loop, is its threshold. The demodulator threshold may be treated as in appendix C given a knowledge of the equivalent closed-loop noise bandwidth B_N. The FM carrier demodulator is shown in figure 4.1-1. The input signal-to-noise ratio is taken as

$$\left.\frac{S_i}{N_i}\right]_{B_{LP}} = \frac{\frac{A^2}{2}}{2\left|\Phi_{n_i}\right|B_{LP}} \qquad (1)$$

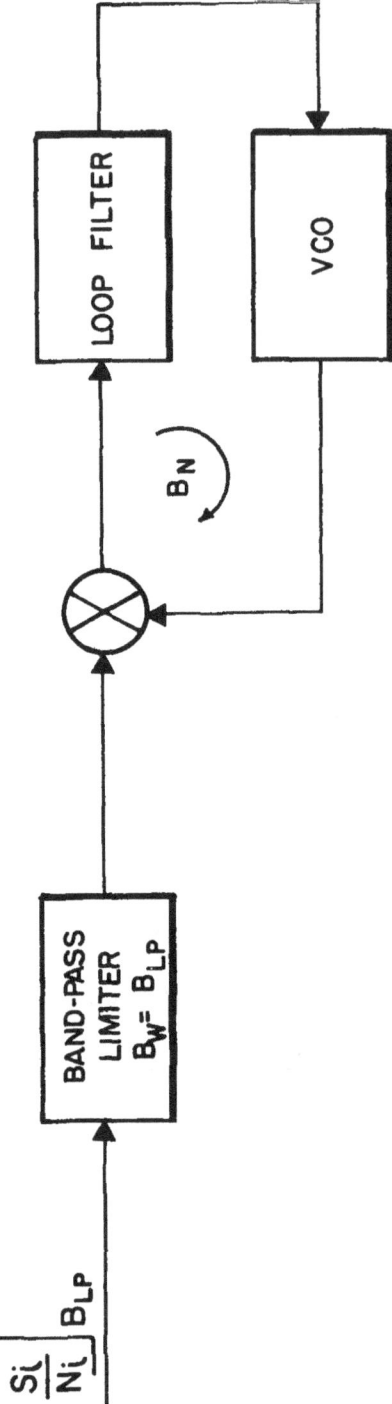

Figure 4.1-1.- The FM carrier channel

where

A = sinusoidal carrier amplitude

$\left|\Phi_{n_i}\right|$ = input noise spectral density magnitude

B_{LP} = bandwidth of the band-pass limiter

At this point we will assume that the carrier channel loop noise bandwidth is much larger than the bandwidth of the band-pass limiter preceding it. Therefore, the threshold of this channel can be treated as in section C.5.

4.2 Television Channel

The performance criterion for the television channel is its output data signal-to-noise ratio. Since the television demodulator is the carrier demodulator, its threshold is treated as stated in section 4.1. Figure 4.2-1 shows the television demodulator.

The input signal-to-noise ratio is taken as

$$\left.\frac{S_i}{N_i}\right]_{B_{LP}} = \frac{\frac{A^2}{2}}{2\left|\Phi_{n_i}\right| B_{LP}} \qquad (1)$$

where

A = sinusoidal carrier amplitude

$\left|\Phi_{n_i}\right|$ = input noise spectral density

B_{LP} = width of the band-pass limiter which feeds the carrier demodulator

The ratio of peak-squared signal to mean-squared noise out of the output low-pass filter is taken from equation E.3.2 (6), page E-11. Thus

$$\left.\frac{S_{TV\ peak}}{N_{TV}}\right]_{B_{TV}} = 3\left[\frac{B_{LP}}{B_{TV}}\right]\left[\frac{\Delta f_{TV}}{B_{TV}}\right]^2 \left[\frac{S_i}{N_i}\right]_{B_{LP}} \qquad (2)$$

where Δf_{TV} is the peak frequency deviation in cycles per second of the television signal on the carrier.

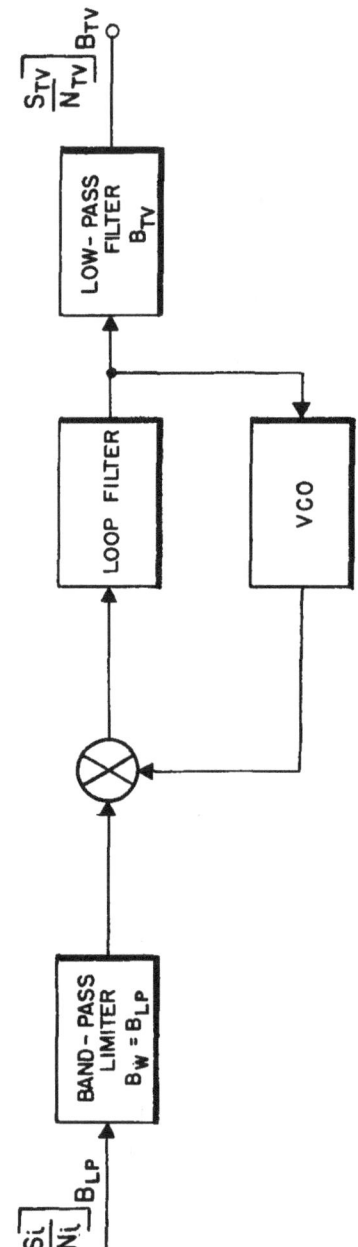

Figure 4.2-1.- The television channel

Utilizing a peak to rms factor K_p, for the television waveform, as in section D.3.2, the mean-squared output signal-to-noise ratio is given as

$$\left.\frac{S_{TV}}{N_{TV}}\right]_{B_{TV}} = K_p^2 \left[\frac{S_{TV\ peak}^2}{N_{TV}}\right] \qquad (3)$$

From equations (2) and (3) the data output signal-to-noise ratio is

$$\left.\frac{S_{TV}}{N_{TV}}\right]_{B_{TV}} = 3K_p^2 \left[\frac{B_{LP}}{B_{TV}}\right]\left[\frac{\Delta f_{TV}}{B_{TV}}\right]^2 \left.\left[\frac{S_i}{N_i}\right]\right]_{B_{LP}}$$

However, since

$$\left[\frac{B_{LP}}{B_{TV}}\right]\left.\left[\frac{S_i}{N_i}\right]\right]_{B_{LP}} \equiv \left.\frac{S_i}{N_i}\right]_{B_{TV}}$$

then

$$\left.\frac{S_{TV}}{N_{TV}}\right]_{B_{TV}} = 3K_p^2 \left[\frac{\Delta f_{TV}}{B_{TV}}\right]^2 \left.\left[\frac{S_i}{N_i}\right]\right]_{B_{TV}} \qquad (4)$$

4.3 PCM Telemetry Channel

The performance criteria for the telemetry channel are its output data signal-to-noise ratio and the telemetry demodulator threshold. Figure 4.3-1 shows the telemetry channel.

The input signal noise ratio is taken as

$$\left.\frac{S_i}{N_i}\right]_{B_{LP}} = \frac{\frac{A^2}{2}}{2\left|\Phi_{n_i}\right| B_{LP}} \qquad (1)$$

where

A = sinusoidal carrier amplitude

$\left|\Phi_{n_i}\right|$ = input noise spectral density

B_{LP} = width of the band-pass limiter which feeds the carrier demodulator

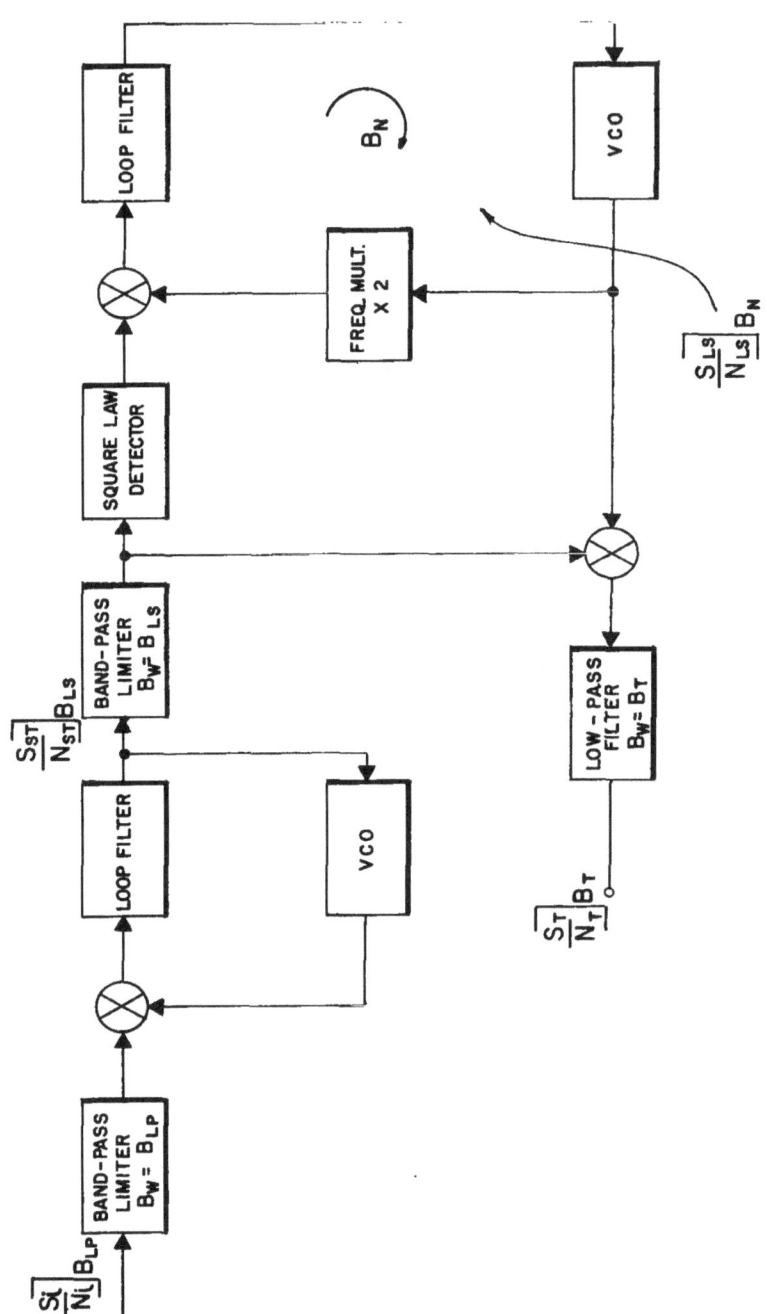

Figure 4.3-1.- PCM telemetry channel

The telemetry subcarrier signal-to-noise ratio at the carrier demodulator output, computed in B_{LS}, the bandwidth of the subcarrier band-pass limiter, is taken from equation E.3.1 (8), page E-10. Thus,

$$\left.\frac{S_{ST}}{N_{ST}}\right]_{B_{LS}} \cong \frac{1}{2}\left[\frac{\Delta f_T}{f_T}\right]^2 \left[\frac{B_{LP}}{B_{LS}}\right]\left[\frac{S_i}{N_i}\right]_{B_{LP}} \quad (2)$$

where

f_T = subcarrier frequency

Δf_T = peak frequency deviation of the subcarrier on the carrier

Equation (2) employs the assumption that

$$\frac{1}{12}\left[\frac{B_{LS}}{f_T}\right]^2 \ll 1 \quad (3)$$

Now the output data signal-to-noise ratio may be obtained. Thus,

$$\left.\frac{S_T}{N_T}\right]_{B_T} = \left[\frac{B_{LS}}{B_T}\right]\left[\frac{S_{ST}}{N_{ST}}\right]_{B_{LS}} \quad (4)$$

where

B_T = bandwidth of the postdetection filter

Using now equations (2) and (4), we obtain:

$$\left.\frac{S_T}{N_T}\right]_{B_T} \cong \frac{1}{2}\left[\frac{\Delta f_T}{f_T}\right]^2 \left[\frac{B_{LP}}{B_T}\right]\left[\frac{S_i}{N_i}\right]_{B_{LP}} \quad (5)$$

However, since

$$\left[\frac{B_{LP}}{B_T}\right]\left[\frac{S_i}{N_i}\right]_{B_{LP}} \equiv \left[\frac{S_i}{N_i}\right]_{B_T}$$

$$\left.\frac{S_T}{N_T}\right]_{B_T} \cong \frac{1}{2}\left[\frac{\Delta f_T}{f_T}\right]^2 \left[\frac{S_i}{N_i}\right]_{B_T} \quad (6)$$

The threshold of the telemetry demodulator has been discussed in section F.3 where the demodulator analysis is given.

4.4 Voice Channel

The performance criteria for the voice channel are the data output signal-to-noise ratio and the voice subcarrier demodulator threshold. The voice channel is shown in figure 4.4-1.

The input signal-to-noise ratio, computed in B_{LP}, the bandwidth of the carrier band-pass limiter is taken as

$$\left.\frac{S_i}{N_i}\right]_{B_{LP}} = \frac{\frac{A^2}{2}}{2\left|\Phi_{n_i}\right|B_{LP}} \tag{1}$$

where

A = sinusoidal carrier amplitude

$\left|\Phi_{n_i}\right|$ = input noise spectral density

The voice subcarrier signal-to-noise ratio at the carrier demodulator output, computed in B_{LS}, the subcarrier band-pass limiter bandwidth, is taken from equation E.3.1 (8), page E-10, as

$$\left.\frac{S_{SV}}{N_{SV}}\right]_{B_{LS}} \simeq \frac{1}{2}\left[\frac{\Delta f_{SV}}{f_{SV}}\right]^2\left[\frac{B_{LP}}{B_{LS}}\right]\left[\frac{S_i}{N_i}\right]_{B_{LP}} \tag{2}$$

where

f_{SV} = voice subcarrier frequency

Δf_{SV} = peak frequency deviation of the subcarrier on the carrier.

Equation (2) uses the assumption that

$$\frac{1}{12}\left[\frac{B_{LS}}{f_{SV}}\right] \ll 1 \tag{3}$$

The peak-squared signal to mean-squared noise at the low-pass filter output, computed in B_v, the low-pass filter bandwidth is taken from equation E.3.2 (6), page E-11, as

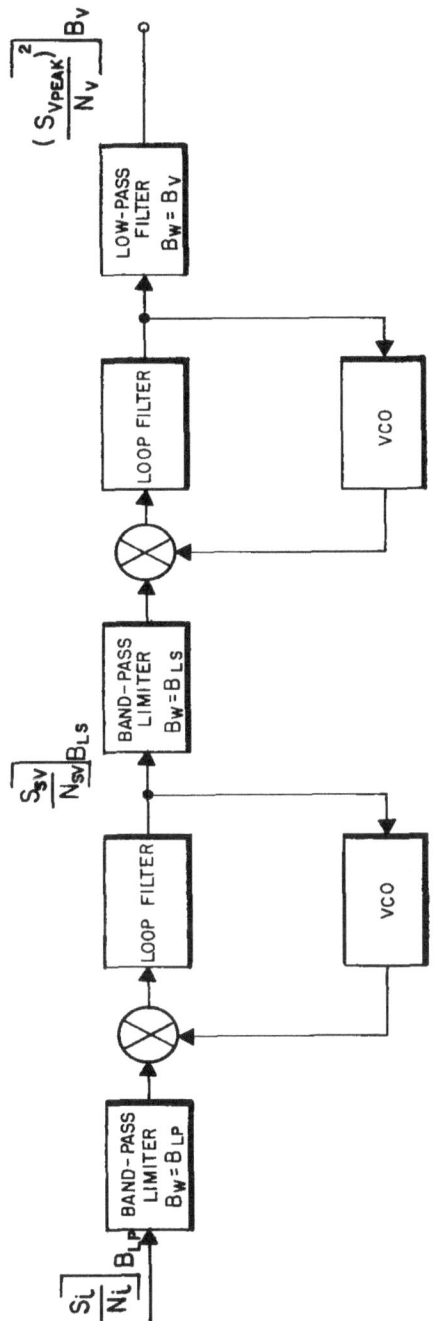

Figure 4.4-1.- The voice channel

$$\left.\frac{(S_{V\ peak})^2}{N_V}\right]_{B_V} = 3\left[\frac{\Delta f_{V\ peak}}{B_V}\right]^2 \left[\frac{B_{LS}}{B_V}\right]\left[\frac{S_{SV}}{N_{SV}}\right]_{B_{LS}} \tag{4}$$

where

$\Delta f_{V\ peak}$ = peak frequency deviation of the voice signal on the voice subcarrier

From equations (2) and (4) the output data signal-to-noise ratio is given as

$$\left.\frac{(S_{V\ peak})^2}{N_V}\right]_{B_V} = \frac{3}{2}\left[\frac{\Delta f_{SV}}{f_{SV}}\right]^2 \left[\frac{\Delta f_{V\ peak}}{B_V}\right]^2 \left[\frac{B_{LP}}{B_V}\right]\left[\frac{S_i}{N_i}\right]_{B_{LP}} \tag{5}$$

However, since

$$\left[\frac{B_{LP}}{B_V}\right]\left[\frac{S_i}{N_i}\right]_{B_{LP}} \equiv \left[\frac{S_i}{N_i}\right]_{B_V}$$

$$\left.\frac{(S_{V\ peak})^2}{N_V}\right]_{B_V} = \frac{3}{2}\left[\frac{\Delta f_{SV}}{f_{SV}}\right]^2 \left[\frac{\Delta f_{V\ peak}}{B_V}\right]^2 \left[\frac{S_i}{N_i}\right]_{B_V} \tag{6}$$

The threshold of the voice demodulator may be treated as in appendix C.

4.5 Biomedical Data Channel

The performance criteria for the biomedical channels are the data output signal-to-noise ratio, the individual biomedical subcarrier demodulator thresholds, and the voice demodulator threshold, since the biomedical subcarriers are frequency multiplexed with the normal voice. Only one biomedical channel is treated in this section, since the general equations are the same for all biomedical channels.

The input signal-to-noise ratio, computed in B_{LP}, the carrier limiter bandwidth is taken as

$$\left.\frac{S_i}{N_i}\right]_{B_{LP}} = \frac{\frac{A^2}{2}}{2\left|\phi_{n_i}\right|B_{LP}}$$

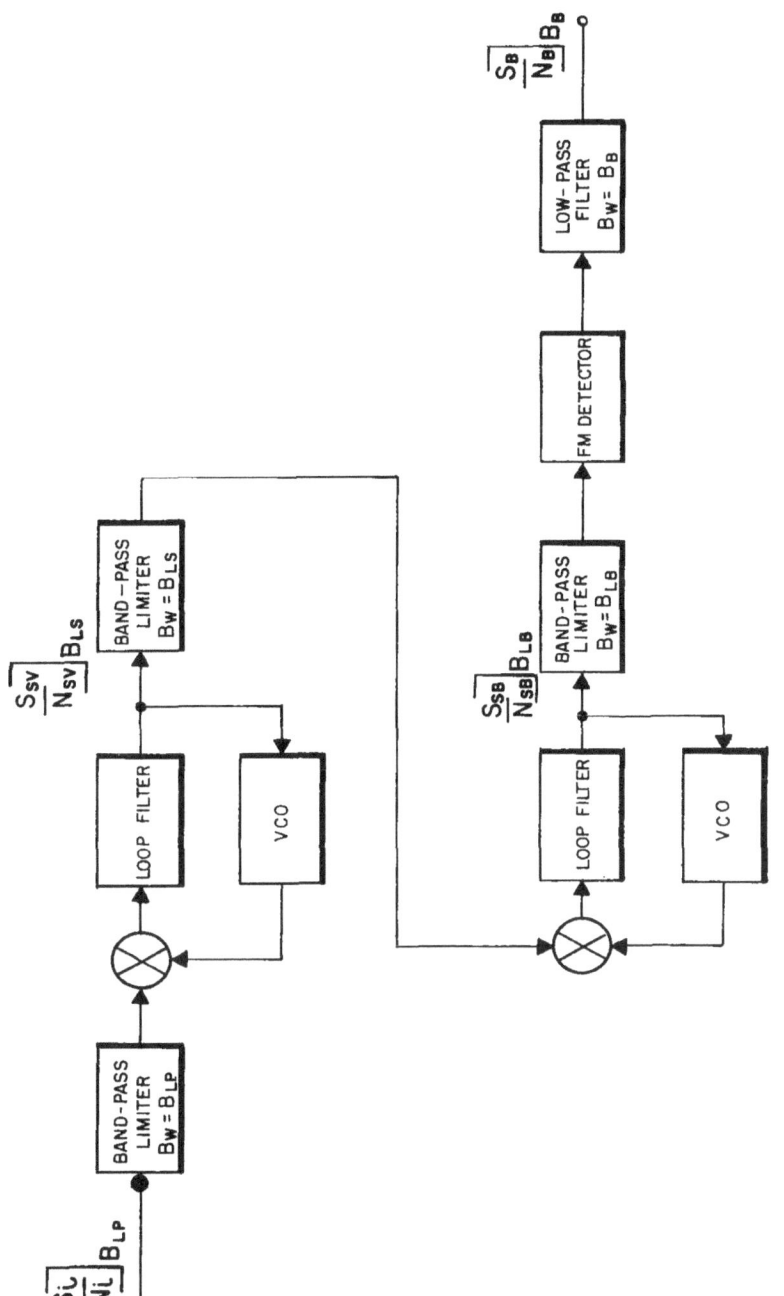

Figure 4.5-1.- The biomedical data channel

where

A = sinusoidal carrier amplitude

$|\Phi_{n_i}|$ = input noise spectral density

The voice subcarrier signal-to-noise ratio at the carrier demodulator output, computed in B_{LS}, the subcarrier limiter bandwidth, is taken from equation E.3.1 (8), page E-10, as

$$\left.\frac{S_{SV}}{N_{SV}}\right]_{B_{LS}} \cong \frac{1}{2}\left[\frac{\Delta f_{SV}}{f_{SV}}\right]^2 \left[\frac{B_{LP}}{B_{LS}}\right] \left[\frac{S_i}{N_i}\right]_{B_{LP}} \qquad (2)$$

where

f_{SV} = voice subcarrier frequency

Δf_{SV} = peak frequency deviation of the subcarrier on the carrier

Equation (2) uses the assumption that

$$\frac{1}{12}\left[\frac{B_{LS}}{f_{SV}}\right]^2 \ll 1 \qquad (3)$$

The biomedical subcarrier signal-to-noise ratio at the output of the voice subcarrier demodulator, computed in B_{LB}, the biomedical subcarrier band-pass limiter bandwidth, is taken from equation E.3.1 (8), page E-10, as

$$\left.\frac{S_{SB}}{N_{SB}}\right]_{B_{LB}} \cong \frac{1}{2}\left[\frac{\Delta f_{SB}}{f_{SB}}\right]^2 \left[\frac{B_{LS}}{B_{LB}}\right] \left[\frac{S_{SV}}{N_{SV}}\right]_{B_{LS}} \qquad (4)$$

where

f_{SB} = biomedical subcarrier frequency

Δf_{SB} = peak frequency deviation of the biomedical subcarrier on the voice subcarrier.

Equation (4) uses the assumptions that the noise spectral density out of the voice subcarrier demodulator may be considered flat across the biomedical subcarrier limiter bandwidth, and that

$$\frac{1}{12}\left[\frac{B_{LB}}{f_{SB}}\right]^2 \ll 1 \qquad (5)$$

The peak-squared signal to mean-squared noise ratio at the output of the low-pass filter, computed in B_B, the low-pass filter bandwidth, is taken from equation E.3.2 (6), page E-11, as

$$\left.\frac{\left(S_{B\ peak}\right)^2}{N_B}\right]_{B_B} = 3\left[\frac{\Delta f_{B\ peak}}{B_B}\right]^2\left[\frac{B_{LB}}{B_B}\right]\left[\frac{S_{SB}}{N_{SB}}\right]_{B_{LB}} \qquad (6)$$

Using a peak-to-rms factor, K_p, for the biomedical data signal, the mean-squared data signal-to-noise ratio at the output is taken from equation E.3.2 (7), page E-11, as

$$\left.\frac{S_B}{N_B}\right]_{B_B} = K_p^2\left[\frac{\left(S_{B\ peak}\right)^2}{N_B}\right]_{B_B} \qquad (7)$$

Combining equations (2), (4), (6), and (7), the output biomedical data signal-to-noise ratio is given as

$$\left.\frac{S_B}{N_B}\right]_{B_B} \cong \frac{3}{4}K_p^2\left[\frac{\Delta f_{B\ peak}}{B_B}\right]^2\left[\frac{\Delta f_{SB}}{f_{SB}}\right]^2\left[\frac{\Delta f_{SV}}{f_{SV}}\right]^2\left[\frac{B_{LP}}{B_B}\right]\left[\frac{S_i}{N_i}\right]_{B_{LP}} \qquad (8)$$

However, since

$$\left[\frac{B_{LP}}{B_B}\right]\left[\frac{S_i}{N_i}\right]_{B_{LP}} \equiv \left[\frac{S_i}{N_i}\right]_{B_B}$$

equation (8) becomes

$$\left.\frac{S_B}{N_B}\right]_{B_B} \cong \frac{3}{4}K_p^2\left[\frac{\Delta f_{B\ peak}}{B_B}\right]^2\left[\frac{\Delta f_{SB}}{f_{SB}}\right]^2\left[\frac{\Delta f_{SV}}{f_{SV}}\right]^2\left[\frac{S_i}{N_i}\right]_{B_B} \qquad (9)$$

The threshold of the biomedical data channels is essentially that of the voice demodulator since voice and the biomedical data subcarriers are frequency multiplexed.

APPENDIX A

ANGLE MODULATION

As explained in volume I of this series, the Apollo lunar communication system employs sinusoidal angle modulation. In particular, both phase modulation (PM) and frequency modulation (FM) are employed. It is the purpose of this appendix to derive usable mathematical models for two angle modulated signals. The first sinusoidal signal is modulated by K sinusoidal subcarriers. The second signal is modulated by the sum of K sinusoidal subcarriers plus a rectangular wave, generated from a pseudo-random ranging code.

A.1 Basic Considerations

A sinusoidal angle modulated signal may be simply represented as

$$s(t) = A \cos \psi(t) \tag{1}$$

where

A = signal amplitude

$\psi(t)$ = time variation of the sinusoid

If a signal function, $f(t)$, is to be incorporated in the signal, two simple methods may be used. For phase modulation, let

$$\psi(t) = \omega_c t + f(t) \quad \text{(PM)} \tag{2}$$

then

$$s_{PM}(t) = A \cos \left[\omega_c t + f(t) \right] \tag{3}$$

If ω_c is taken as the unmodulated frequency of the sinusoidal signal (or carrier), then it is seen that the signal function $f(t)$ appears directly in the signal phase. For frequency modulation, let

$$\psi(t) = \omega_c t + \int_t f(t) \, dt \quad \text{(FM)} \tag{4}$$

then

$$s_{FM}(t) = A \cos \left[\omega_c t + \int_t f(t) \, dt \right] \tag{5}$$

Now, the instantaneous frequency of the sinusoidal signal may be defined as the time derivative of the angle $\psi(t)$

$$F(t) = \dot{\psi}(t) = \omega_c + f(t) \tag{6}$$

It is seen that in FM, the signal function $f(t)$ appears directly in the signal frequency.

Based on the preceding equations, one notation may be used to represent either PM or FM modulated signals.

$$s(t) \triangleq A \cos\left[\omega_c t + \varphi_s(t)\right] \tag{7}$$

$\varphi_s(t)$ is the equivalent phase modulation of the signal. In terms of a signal function $f(t)$,

$$\text{for PM:} \quad \varphi_s(t) = f(t) \tag{8}$$

$$\text{for FM:} \quad \varphi_s(t) = \int_t f(t)\, dt \tag{9}$$

Most useful periodic signal functions can be represented as terminated Fourier series in the form.

$$f(t) = a_o + \sum_{i=1}^{K} \left[a_i \cos \omega_i t + b_i \sin \omega_i t\right] \tag{10}$$

$$f(t) = c_o + \sum_{i=1}^{K} c_i \cos\left[\omega_i t + \theta_i\right] \tag{11}$$

or

$$f(t) = d_o + \sum_{i=1}^{K} d_i \sin\left[\omega_i t + \varphi_i\right] \tag{12}$$

where

$$\left.\begin{array}{l} c_o \text{ and } d_o \text{ are constants} \\[6pt] c_i = d_i = \sqrt{a_i^2 + b_i^2} \\[6pt] \theta_i = -\tan^{-1}\left[\dfrac{b_i}{a_i}\right] \\[6pt] \varphi_i = \tan^{-1}\left[\dfrac{a_i}{b_i}\right] \end{array}\right\} \quad (13)$$

For a given signal function, c_o and d_o are non-time varying and convey no information. They will be deleted for simplification. Then the signal function may be defined

$$f(t) \triangleq \sum_{i=1}^{K} f_i \begin{Bmatrix} \sin \\ \cos \end{Bmatrix} \left[\omega_i t + \begin{Bmatrix} \varphi_i \\ \theta_i \end{Bmatrix} \right]$$

$$f_i \triangleq \begin{Bmatrix} d_i \\ c_i \end{Bmatrix} \quad (14)$$

where the brackets indicate "either/or".

For PM we may define

$$\varphi_s(t) = \sum_{i=1}^{K} \Delta\varphi_i \sin\left[\omega_i t + \varphi_i\right] \quad (15)$$

where

$\Delta\varphi_i$ = peak phase deviation due to the ith component.

It is seen that the definition of equation (15) places no restrictions on φ_i. For the case where $\varphi_s(t)$ represents K subcarriers, the subcarriers themselves may be angle modulated, in which case φ_i is a function of time.

For FM, we may define the frequency function

$$\dot{\varphi}_s(t) = \sum_{i=1}^{K} \Delta\omega_i \cos\left[\omega_i t + \theta_i\right] \tag{16}$$

where

$\Delta\omega_i$ = peak radian frequency deviation due to the ith component

It is seen that the definition of equation (16) places no restriction on θ_i. For the case where $\dot{\varphi}_s(t)$ represents K subcarriers, the subcarriers themselves may be angle modulated, in which case θ_i is a function of time. It is difficult to determine, analytically, the equivalent phase modulation $\varphi_s(t)$ when θ_i is a function of time, due to the difficulty of integrating the frequency function. However, this difficulty does not invalidate the concept of representing FM by an equivalent phase modulation.

A.2 The Carrier with K Subcarriers

This signal may be represented as

$$s(t) = A \begin{Bmatrix} \sin \\ \cos \end{Bmatrix} \left[\omega_c t + \varphi_s(t)\right] \tag{1}$$

We will examine first, the PM case. Let

$$\varphi_s(t) = \sum_{i=1}^{K} \Delta\varphi_i \sin\left[\omega_i t + \varphi_i\right] \tag{2}$$

Then

$$s(t) = A \begin{Bmatrix} \sin \\ \cos \end{Bmatrix} \left[\omega_c t + \sum_{i=1}^{K} \Delta\varphi_i \sin\left(\omega_i t + \varphi_i\right)\right] \tag{3}$$

Equation (3) has previously been treated by Giacolleto (ref.1). The result obtained is

$$s(t) \equiv A \sum_{n_1=-\infty}^{\infty} \cdots \sum_{n_k=-\infty}^{\infty} \prod_{i=1}^{K} \left[J_{n_i}(\Delta\varphi_i) \right] \begin{Bmatrix} \sin \\ \cos \end{Bmatrix} \left[\omega_c t + \sum_{i=1}^{K} n_i(\omega_i t + \varphi_i) \right] \quad (4)$$

The Bessel function expansions and summing processes leading to equation (4) are unaffected by the time behavior of φ_i. Therefore, equation (4) is valid for angle modulated subcarriers.

The residual carrier is defined as that term remaining after modulation at the frequency ω_c of the unmodulated carrier. Observation of equation (4) shows that the residual carrier term of frequency ω_c is that term for which all n_i are identically zero.

$$n_i \equiv 0 \quad (5)$$

Then

$$s_c(t) \equiv A \prod_{i=1}^{K} J_0(\Delta\varphi_i) \begin{Bmatrix} \sin \\ \cos \end{Bmatrix} \omega_c t \quad (6)$$

The FM case may be treated exactly only for unmodulated subcarriers, where θ_i is constant. The case may be treated approximately if the greatest frequency component in θ_i is much less than the subcarrier frequency ω_i. The equivalent phase modulation is

$$\varphi_s(t) = \int_t \left[\sum_{i=1}^{K} \Delta\omega_i \cos(\omega_i t + \theta_i) \right] dt \quad (7)$$

$$\varphi_s(t) \cong \sum_{i=1}^{K} \frac{\Delta\omega_i}{\omega_i} \sin(\omega_i t + \theta_i) \quad (8)$$

The equivalent peak phase deviation is seen to be

$$\Delta\varphi_i \cong \frac{\Delta\omega_i}{\omega_i} \tag{9}$$

Then, the FM signal may be represented as

$$s(t) \cong A \begin{Bmatrix} \sin \\ \cos \end{Bmatrix} \left[\omega_c t + \sum_{i=1}^{K} \frac{\Delta\omega_i}{\omega_i} \sin(\omega_i t + \theta_i) \right] \tag{10}$$

Equation (10) holds only with the restriction on θ_i, previously mentioned. The Giacoletto (ref. 1) expansion is then

$$s(t) \cong A \sum_{n_i=-\infty}^{\infty} \cdots \sum_{n_k=-\infty}^{\infty} \prod_{i=1}^{K} \left[J_{n_i}\left(\frac{\Delta\omega_i}{\omega_i}\right) \right] \begin{Bmatrix} \sin \\ \cos \end{Bmatrix} \left[\omega_c t + \sum_{i=1}^{K} n_i(\omega_i t + \theta_i) \right] \tag{11}$$

As in equation (6) the residual carrier is obtained as

$$s_c(t) \cong A \prod_{i=1}^{K} \left[J_0\left(\frac{\Delta\omega_i}{\omega_i}\right) \right] \begin{Bmatrix} \sin \\ \cos \end{Bmatrix} \omega_c t \tag{12}$$

A.3 The Carrier with K Subcarriers and Range Code

This section extends the Giacoletto (ref. 1) expansion for a carrier which is phase modulated by K sinusoidal subcarriers plus a square waveform, representing a pseudo-random ranging code.

The modulated signal is represented as

$$s(t) = A \begin{Bmatrix} \sin \\ \cos \end{Bmatrix} \left[\omega_c t + \Delta\varphi_r c_t(t) + \sum_{i=1}^{K} \Delta\varphi_i \sin(\omega_i t + \varphi_i) \right] \tag{1}$$

where $\Delta\varphi_r$ is the peak phase deviation of the range code on the carrier, and $c_t(t)$ is a code waveform, having only the values ±1. It is assumed that the code makes instantaneous transitions between the + and − 1 states.

Equation (1) may be expanded as

$$s(t) = A \left(\begin{Bmatrix} \sin[\Delta\varphi_r c_t(t)] \\ \cos[\Delta\varphi_r c_t(t)] \end{Bmatrix} \cos\left[\omega_c t + \sum_{i=1}^{K} \Delta\varphi_i \sin(\omega_i t + \varphi_i)\right] \right.$$

$$\left. + \begin{Bmatrix} \cos[\Delta\varphi_r c_t(t)] \\ -\sin[\Delta\varphi_r c_t(t)] \end{Bmatrix} \sin\left[\omega_c t + \sum_{i=1}^{K} \Delta\varphi_i \sin(\omega_i t + \varphi_i)\right] \right) \quad (2)$$

Now

$$\left[\sin \Delta\varphi_r c_t(t)\right] = \begin{Bmatrix} \sin(\Delta\varphi_r); \ c_t(t) = +1 \\ -\sin(\Delta\varphi_r); \ c_t(t) = -1 \end{Bmatrix} = c_t(t) \sin(\Delta\varphi_r) \quad (3)$$

and

$$\cos\left[\Delta\varphi_r c_t(t)\right] = \begin{Bmatrix} \cos(\Delta\varphi_r); \ c_t(t) = +1 \\ \cos(\Delta\varphi_r); \ c_t(t) = -1 \end{Bmatrix} = \cos(\Delta\varphi_r) \quad (4)$$

Combining equations (2), (3), and (4), we obtain

$$s(t) = A \left(\begin{Bmatrix} c_t(t) \sin(\Delta\varphi_r) \\ \cos(\Delta\varphi_r) \end{Bmatrix} \cos\left[\omega_c t + \sum_{i=1}^{K} \Delta\varphi_i \sin(\omega_i t + \varphi_i)\right] \right.$$

$$\left. + \begin{Bmatrix} \cos(\Delta\varphi_r) \\ -c_t(t) \sin(\Delta\varphi_r) \end{Bmatrix} \sin\left[\omega_c t + \sum_{i=1}^{K} \Delta\varphi_i \sin(\omega_i t + \varphi_i)\right] \right) \quad (5)$$

Equation (5) contains two expressions which may be treated with the Giocoletto (ref. 1) expansion to give

$$s(t) = A \left(\begin{Bmatrix} c_t(t) \sin(\Delta\varphi_r) \\ \cos \Delta\varphi_r \end{Bmatrix} \sum_{n_1=-\infty}^{\infty} \cdots \sum_{n_k=-\infty}^{\infty} \prod_{i=1}^{K} \left[J_{n_i}(\Delta\varphi_i) \right] \cos \left[\omega_c t + \sum_{i=1}^{K} n_i(\omega_i t + \varphi_i) \right] + \begin{Bmatrix} \cos \Delta\varphi_r \\ -c_t(t) \sin \Delta\varphi_r \end{Bmatrix} \sum_{n_1=-\infty}^{\infty} \cdots \sum_{n_k=-\infty}^{\infty} \prod_{i=1}^{K} \left[J_{n_i}(\Delta\varphi_i) \right] \sin \left[\omega_c t + \sum_{i=1}^{K} n_i(\omega_i t + \varphi_i) \right] \right) \quad (6)$$

Equation (6) will be used subsequently in appendices D and G to obtain explicit expressions for detected subcarriers and range code. As in equation A.2 (12), page A-6, the residual carrier term may be obtained as

$$s_c(t) = A \cos(\Delta\varphi_r) \prod_{i=1}^{K} \left[J_0(\Delta\varphi_i) \right] \begin{Bmatrix} \sin \\ \cos \end{Bmatrix} \omega_c t \quad (7)$$

APPENDIX B

NOISE

This appendix sets down all the relationships for noise which are used in the remainder of the document. The governing assumption shall be that all noise processes encountered at the inputs of the various systems and subsystems shall be considered as characterized by Gaussian statistics. That is, all input noise wave forms will be taken as stationary, random, Gaussian processes. Since the treatments of Gaussian processes and combinations of deterministic signals summed with Gaussian noise are well documented, only the pertinent results will be set down here, along with references to the original treatments.

B.1 The Narrow-Band Gaussian Random Process

A Gaussian noise process which has a spectral width, Δf, much less than its center frequency, f_c, can be expressed in a very meaningful, useful form. A sample function of the process will be represented as $n(t)$. The sample function $n(t)$ may be expressed as the difference of two components in phase quadrature as in reference 2.

$$n(t) = x(t) \cos \omega_c t - y(t) \sin \omega_c t \qquad (1)$$

where ω_c is the radian center frequency of the specrum of $n(t)$. Bennett (ref. 2) shows that $x(t)$ and $y(t)$ are sample functions of independent Gaussian processes. Davenport and Root (ref. 3) show that the possible values of $x(t)$ and $y(t)$ are determined by Gaussian variables x and y, which have expected values, or means, of zero, and whose variances are related to the variance of the original sample function $n(t)$ by

$$\sigma_x^2 = \sigma_y^2 = \sigma_n^2 \qquad (2)$$

Moreover, if the narrow-band process is characterized by a noise spectral density, $\Phi_n(\omega)$ watts per cycle of bandwidth, the spectral densities of the x and y components are related as

$$\Phi_x(\omega) = \Phi_y(\omega) = \Phi_n(\omega + \omega_c) + \Phi_n(\omega - \omega_c) \qquad (3)$$

where equation (3) holds regardless of limitations of bandwidth of the original process (ref. 2).

A transformation from rectangular to polar coordinates yields

$$n(t) = p(t) \cos\left[\omega_c t + \varphi(t)\right] \tag{4}$$

where $p(t)$ is identified as the "envelope", and $\varphi(t)$ the "phase" of $n(t)$. Both $p(t)$ and $\varphi(t)$ are sample functions of random processes which are not independent (ref. 3). The density of the random variable p is Rayleigh, while the density of the random variable φ is uniform in the range $(0, 2\pi)$.

It should be noted that since spectral densities of time functions are either Fourier transforms of autocorrelation functions or products of Fourier transforms with conjugate Fourier transforms, and since the original time functions are real, then the spectral densities are two-sided; that is, the spectral densities exist for positive and negative real frequencies.

A special case of the Gaussian narrow-band process occurs when the spectral density, $\Phi_n(\omega)$, may be considered, within the narrow frequency region of interest, to be a constant. That is, for a center frequency, ω_c, it may be assumed that

$$\Phi_n(\omega) = |\Phi_n| \; ; \quad \bigl||\omega| - \omega_c\bigr| \leq \frac{\Delta\omega}{2}$$

$$= 0; \text{ all other } \omega \tag{5}$$

where $|\Phi_n|$ is a constant. The spectrum of $\Phi_n(\omega)$ consists of two square blocks of intensity $|\Phi_n|$ and width $\Delta\omega$, centered on $+\omega_c$ and $-\omega_c$, respectively, as shown in figure B.1-1.

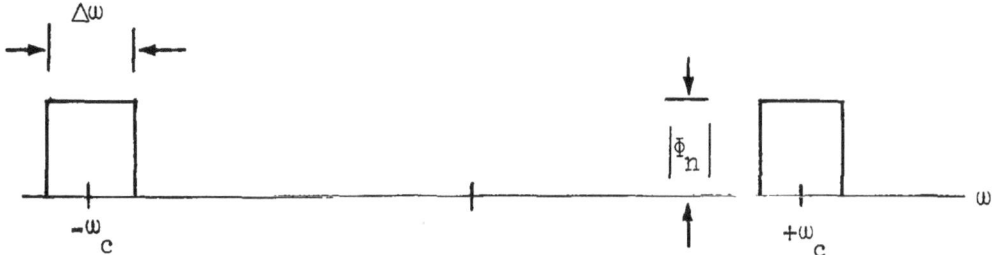

Figure B.1-1.- Input noise spectrum

Then the low frequency components have spectral densities

$$\Phi_x(\omega) = \Phi_y(\omega) = 2|\Phi_n|; \quad |\omega| \leq \frac{\Delta\omega}{2}$$

$$= 0 \; ; \text{ all other } \omega \tag{6}$$

This special case is useful in the treatment of thermally generated noise processes.

B.2 Angle Modulated Carrier Plus Noise

This section treats the sum of an angle modulated (PM or FM) carrier plus Gaussian narrow-band noise. It is assumed that the carrier is centered in the spectral density of the noise. The signal is represented as

$$s(t) = A \cos \left[\omega_c t + \varphi_s(t) \right] \tag{1}$$

where

A = carrier amplitude

ω_c = carrier radian frequency

$\varphi_s(t)$ = carrier "equivalent" phase modulation as treated in appendix A

The sum of signal plus noise is written as

$$s(t) + n(t) = A \cos \left[\omega_c t + \varphi_s(t) \right] + n(t) \tag{2}$$

where

$n(t)$ = a sample function of a narrow-band Gaussian process

$$s(t) + n(t) = A \cos \left[\omega_c t + \varphi_s(t) \right] + x(t) \cos \omega_c t - y(t) \sin \omega_c t \tag{3}$$

$$s(t) + n(t) = \left[A \cos \varphi_s(t) + x(t) \right] \cos \omega_c t - \left[A \sin \varphi_s(t) + y(t) \right] \sin \omega_c t \tag{4}$$

A transformation to polar coordinates gives

$$s(t) + n(t) = A(t) \cos \left[\omega_c t + \psi(t) \right] \tag{5}$$

where

$$A(t) = \left\{ \left[A \cos \varphi_s(t) + x(t)\right]^2 + \left[A \sin \varphi_s(t) + y(t)\right]^2 \right\}^{\frac{1}{2}} \quad (6)$$

$$\sin \psi(t) = \frac{A \sin \varphi_s(t) + y(t)}{A(t)} \quad (7)$$

$$\cos \psi(t) = \frac{A \cos \varphi_s(t) + x(t)}{A(t)} \quad (8)$$

$$\psi(t) = \arctan \left[\frac{A \sin \varphi_s(t) + y(t)}{A \cos \varphi_s(t) + x(t)}\right] \quad (9)$$

Equation (5) shows that the effect of summing Gaussian noise with an angle modulated carrier can be interpreted as that of producing a signal which is simultaneously amplitude and angle modulated. It is seen that the amplitude or envelope function $A(t)$ is not negative.

For the special case of an angle modulated signal embedded in narrow-band white Gaussian noise with a relatively high ratio of carrier-to-noise, Bennett (ref. 2) has shown that equation (3) and (5) may be well approximated by

$$s(t) + n(t) \cong A \cos \left[\omega_c t + \varphi_s(t) + \frac{y(t)}{A}\right]; \quad A^2 \gg \overline{y^2(t)} \quad (10)$$

Thus, in the low-noise case, additive white band-limited Gaussian noise may be considered to add a separate "phase" jitter to an angle modulated signal. Where the approximation holds, the phase noise may be considered to be a sample function of a Gaussian process. Thus, for the special case, the spectral density, $\Phi_\varphi(\omega)$, of the phase noise may be expressed as

$$\Phi_\varphi(\omega) = \frac{\Phi_y(\omega)}{A^2} \equiv \frac{2|\Phi_n|}{A^2}; \quad |\omega| < \frac{\Delta\omega}{2} \quad (11)$$

or

$$\Phi_\varphi(\omega) = \frac{|\Phi_n|}{S}; \quad \omega < \frac{\Delta\omega}{2} \quad (12)$$

where S is the angle modulated carrier power.

B.3 Transmission of Signal Plus Noise Through a Perfect Band-pass Limiter

The treatment of passing a carrier plus Gaussian noise through a limiter is complex and has been performed by Davenport (ref. 4), Middleton (ref. 5), and others. Section H.3 summarizes some results of Davenport's analysis.

The treatment here will be more intuitive, drawing on Davenport's results as needed. Figure B.3-1 shows the block model.

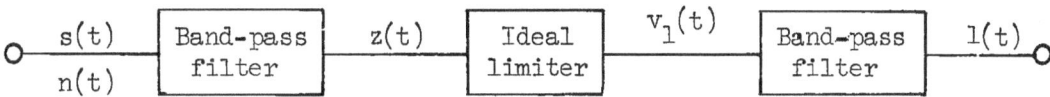

Figure B.3-1.- Limiter model.

The sum of input signal plus noise is taken in polar form from equation B.2 (5), page B-3, as

$$s(t) + n(t) = A(t) \cos \left[\omega_c t + \psi(t) \right] \tag{1}$$

The ideal characteristics of the limiter output $v_1(t)$ are represented as

$$\left. \begin{array}{l} v_1(t) = +V_L; \; z(t) > 0 \\ = -V_L; \; z(t) < 0 \\ = 0; \; z(t) = 0 \end{array} \right\}; \; z(t) = s(t) + n(t) \tag{2}$$

where

V_L = voltage limiting level

The output filter of the limiter is assumed to be a perfect band-pass filter having a flat amplitude transmission characteristic, square frequency cut-off characteristic, and flat phase transmission characteristic across the pass-band. The transmission constant is arbitrarily taken as unity. The filter pass-band is assumed wide enough to pass all zonal energy associated with the carrier frequency ω_c and narrow enough to reject all other zones. From Davenport (ref. 4), the total power P_L out of the band-pass filter is taken as

$$P_L = 8\left[\frac{V_L}{\pi}\right]^2 \tag{3}$$

The output waveform $l(t)$ is taken as

$$l(t) = 4\frac{V_L}{\pi}\cos\left[\omega_c t + \psi(t)\right] \tag{4}$$

As in equation B.2 (10), page B-4, for the special case of narrow-band, white Gaussian noise and high carrier-to-noise ratio, equation (4) may be approximated as

$$l(t) \cong 4\frac{V_L}{\pi}\cos\left[\omega_c t + \varphi_s(t) + \frac{y(t)}{A}\right] ; \quad A^2 \gg \overline{y^2(t)} \tag{5}$$

For this special case, the phase noise spectral density is given as

$$\Phi_\varphi(\omega) = \frac{|\Phi_n|}{S} ; \quad |\omega| < \frac{\Delta\omega}{2} \tag{6}$$

where

$|\Phi_n|$ = constant value of the white noise spectral density into the limiter

S = limiter input signal power

$\Delta\omega$ = bandwidth of the input noise spectral density

B.4 Transmission of Signal Plus Noise Through a Perfect Product Device

B.4.1 A Nonprelimited Product Detector

The product detector of figure B.4.1-1 is fed an angle modulated signal plus narrow-band white Gaussian noise. The reference signal has negative sine phase with respect to the angle modulated carrier.

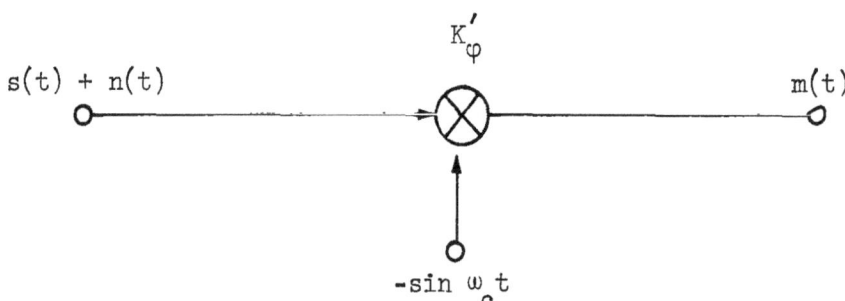

Figure B.4.1-1.- Nonprelimited product detector

The product detector is assumed to have some gain constant, K'_φ, and to reject all except the "d.c." or difference terms of the product. Then

$$m(t) = -K'_\varphi \sin \omega_c t \left[s(t) + n(t)\right] \tag{1}$$

$$m(t) = -K'_\varphi \sin \omega_c t \left\{A \cos\left[\omega_c t + \varphi_s(t)\right] + x(t) \cos \omega_c t - y(t) \sin \omega_c t\right\} \tag{2}$$

The difference terms at the output are

$$m(t) = K'_\varphi \frac{A}{2} \sin \varphi_s(t) + K'_\varphi \frac{y(t)}{2} \tag{3}$$

It is seen that the output function is easily separable into a signal component, $s_m(t)$, and a noise component, $n_m(t)$, as

$$s_m(t) = K'_\varphi \frac{A}{2} \sin \varphi_s(t) \tag{4}$$

$$n_m(t) = K'_\varphi \frac{y(t)}{2} \tag{5}$$

The output noise spectral density is given as

$$\Phi_{n_m}(\omega) = \frac{K'^2_\varphi}{4} \Phi_y(\omega) \tag{6}$$

$$\Phi_{n_m}(\omega) = \frac{K'^2_\varphi}{2} \left|\Phi_n\right| \quad ; \quad \omega < \frac{\Delta\omega}{2}$$

$$\qquad\qquad = 0 \qquad ; \text{ all other } \omega \tag{7}$$

B.4.2 A Prelimited Product Detector

The band-pass limiter of figure B.4.2-1 is fed an angle modulated signal plus narrow-band Guassian noise and drives a product detector, having gain constant $K'\varphi$.

The reference signal has negative sine phase.

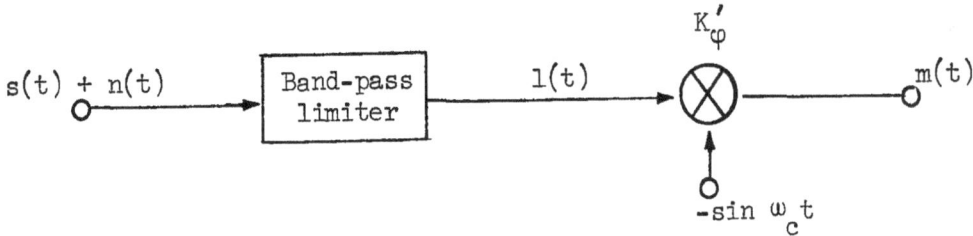

Figure B.4.2-1.- Prelimited product detector

From equation B.3 (4), page B-6,

$$l(t) = 4 \frac{V_L}{\pi} \cos\left[\omega_c t + \psi(t)\right] \tag{1}$$

where V_L is voltage limiting level. Equation B.2 (9), page B-4, describes $\psi(t)$. Then

$$m(t) = -K'_\varphi \sin \omega_c t \, l(t) \tag{2}$$

$$m(t) = -4 \frac{V_L K'_\varphi}{\pi} \sin \omega_c t \cos\left[\omega_c t + \psi(t)\right] \tag{3}$$

The "d.c." or difference term is given as

$$m(t) = \frac{2 V_L K'_\varphi}{\pi} \sin \psi(t) \tag{4}$$

From equation B.2 (7) page B-5

$$m(t) = \frac{2 V_L K'_\varphi}{\pi} \left[\frac{A \sin \varphi_s(t) + y(t)}{A(t)}\right] \tag{5}$$

For the special case of relatively high signal-to-noise ratio into the limiter, the approximation holds that

$$A(t) \cong A \tag{6}$$

and

$$m(t) \cong \frac{2V_L K'_\varphi}{\pi}\left[\sin \varphi_s(t) + \frac{y(t)}{A}\right] \quad (7)$$

Where the approximation holds, the product detector output is separable into signal and noise components

$$s_m(t) = \frac{2V_L K'_\varphi}{\pi} \sin \varphi_s(t) \quad (8)$$

$$n_m(t) = \frac{2V_L K'_\varphi}{\pi A} y(t) \quad (9)$$

and the output noise spectral density is given as

$$\Phi_{n_m}(\omega) = 8\left[\frac{V_L K'_\varphi}{\pi A}\right]^2 |\Phi_n| \quad ; \quad |\omega| < \frac{\Delta\omega}{2} \quad (10)$$

$$= 0 \quad ; \text{ all other } \omega$$

B.4.3 A Nonprelimited Product Mixer

For the device shown in figure B.4.3-1, several conditions are stated. The input to the product device, having gain constant K'_φ, is an angle modulated signal and narrow-band Gaussian noise, centered on a radian frequency ω_c. The reference signal has negative sine phase, arbitrarily, and is of a frequency ω_d such that the output spectra of the sum and difference terms do not overlap. The ideal band-pass filter transmits all energy associated with the difference terms, arbitrarily, and rejects the sum terms.

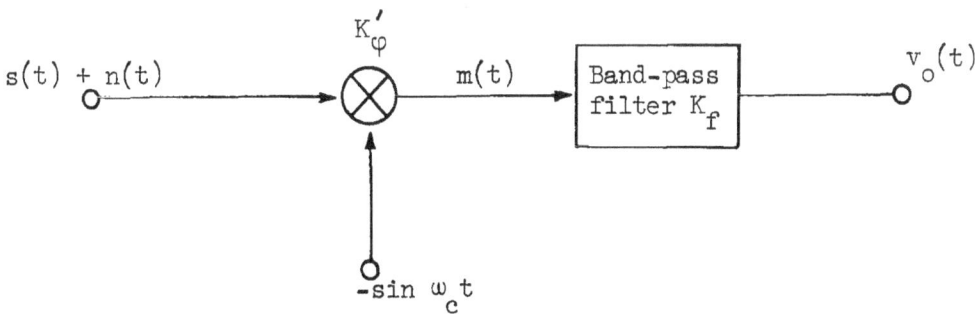

Figure B.4.3-1.- Nonprelimited product mixer

The band-pass filter has an arbitrary gain constant K_f.

$$v_o(t) = K_f \big[m(t)\big]_{\text{difference terms}} \tag{1}$$

$$m(t) = -K'_\varphi \sin \omega_d t \, \big[s(t) + n(t)\big] \tag{2}$$

$$m(t) = -K'_\varphi \sin \omega_d t \left\{ A \cos\big[\omega_c t + \varphi_s(t)\big] + x(t) \cos \omega_c t - y(t) \sin \omega_c t \right\} \tag{3}$$

$$v_o(t) = \frac{K'_\varphi K_f}{2} \left\{ A \sin\big[(\omega_c - \omega_d)t + \varphi_s(t)\big] + x(t) \sin(\omega_c - \omega_d)t + y(t) \cos(\omega_c - \omega_d)t \right\} \tag{4}$$

It is seen that the output function consists of a signal component and a noise component.

$$s_o(t) = \frac{K'_\varphi K_f}{2} A \sin\big[(\omega_c - \omega_d)t + \varphi_s(t)\big] \tag{5}$$

$$n_o(t) = \frac{K'_\varphi K_f}{2} \left[x(t) \sin(\omega_c - \omega_d)t + y(t) \cos(\omega_c - \omega_d)t \right] \tag{6}$$

The output signal and noise functions have the same form as the input functions, being merely multiplied by constants and translated in frequency. Equation (6) can be transformed to the usual recognizable form of equation B.1 (1), page B-1, by a suitable redefinition of variables. For the special case of white Gaussian noise at the input, the output noise spectral density is given as

$$\Phi_{n_o}(\omega) = \left[\frac{K'_\varphi K_f}{2}\right]^2 |\Phi_n| \; ; \; \left||\omega| - \omega_c + \omega_d\right| \leq \frac{\Delta\omega}{2} \tag{7}$$

$$= 0 \quad ; \text{ all other } \omega$$

It is interesting to compute the ratio of output carrier power S_o to the magnitude of the output noise spectral density for the white Gaussian case

$$\frac{S_o}{|\Phi_{n_o}|} = \frac{\left[\frac{K'_\varphi K_f}{2}\right]^2 \frac{A^2}{2}}{\left[\frac{K'_\varphi K_f}{2}\right]^2 |\Phi_n|} \tag{8}$$

$$\frac{S_o}{|\Phi_{n_o}|} \equiv \frac{S}{|\Phi_n|} \tag{9}$$

where

S = input carrier power

$|\Phi_n|$ = magnitude of the input noise spectral density

It is seen that the ratio is constant through a perfect product mixer and ideal band-pass filter, regardless of transmission gain constants.

APPENDIX C

PHASE-LOCKED LOOP THEORY

C.1 A Physical Approach to the Phase-locked Loop

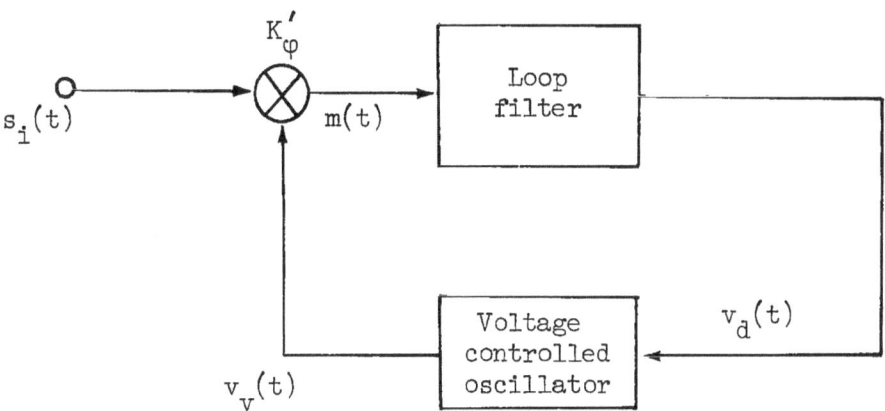

Figure C.1-1.- Physical loop model

Figure C.1-1 shows a model of the phase-locked loop which is essentially a closed loop feedback multiplier. The device consists of a multiplier, having multiplication constant K'_φ, a loop filter, of the low-pass type, and a voltage controlled oscillator (VCO). The physical operation of the loop may best be explored by assuming a noiseless input signal, $s_i(t)$.

The input signal is taken as an angle modulated sinusoid of amplitude A, frequency ω_c, having "effective" phase modulation $\varphi_i(t)$.

$$s_i(t) = A \cos\left[\omega_c t + \varphi_i(t)\right] \qquad (1)$$

Without loss of generality, the VCO may be considered to produce an angle modulated sinusoid of amplitude A_v, frequency ω_c, having an effective phase modulation $\varphi_o(t)$, and having negative sine phase with respect to the input signal. Assuming negative sine phase and a

C-1

frequency ω_c exactly equal to the input signal frequency is not a restriction on generality, since a term linear in t and a constant may be taken in the otherwise unspecified output phase $\varphi_o(t)$.

$$v_V(t) \triangleq -A_V \sin\left[\omega_c t + \varphi_o(t)\right] \qquad (2)$$

The output of the multiplier $m(t)$ is

$$m(t) = s_i(t)v_V(t) = -AA_V K'_\varphi \cos\left[\omega_c t + \varphi_i(t)\right] \sin\left[\omega_c t + \varphi_o(t)\right] \qquad (3)$$

$$m(t) = \frac{-AA_V K'_\varphi}{2}\left\{\sin\left[2\omega_c t + \varphi_i(t) + \varphi_o(t)\right] - \sin\left[\varphi_i(t) - \varphi_o(t)\right]\right\} \qquad (4)$$

It may now be assumed that the multiplier is fully balanced so that the double frequency term is rejected. Also, the low-pass filter will not pass double frequency components. In any event, the desired low-frequency multiplier output is given as

$$m(t) = \frac{AA_V K'_\varphi}{2} \sin\left[\varphi_i(t) - \varphi_o(t)\right] \qquad (5)$$

The VCO driving function $v_d(t)$ is given as the time convolution of the multiplier signal with the impulse response function $h(t)$ of the low-pass filter.

Assuming zero initial conditions at time zero

$$v_d(t) = \int_0^\infty m(t - \tau)h(\tau)\, d\tau \qquad (6)$$

$$v_d(t) = \frac{AA_V K'_\varphi}{2} \int_0^\infty h(\tau) \sin\left[\varphi_i(t - \tau) - \varphi_o(t - \tau)\right] d\tau \qquad (7)$$

The output phase $\varphi_o(t)$ of the VCO is proportional to the time integral of the driving signal. The proportionality constant is K_V

$$\varphi_o(t) = K_V \int_0^t v_d(t)\, dt \tag{8}$$

Thus

$$\varphi_o(t) = \frac{A A_V K'_\varphi K_V}{2} \int_0^t \int_0^\infty h(\tau) \sin\left[\varphi_i(t-\tau) - \varphi_o(t-\tau)\right] d\tau\, dt \tag{9}$$

Equation (9) is the exact nonlinear integral equation giving the VCO phase response to an input phase function. Equation (9) shows that the output phase $\varphi_o(t)$ responds to the input phase $\varphi_i(t)$, but the equation does not give one an intuitive "feel" for the manner in which the loop responds.

Suppose, through some unspecified means, over some interval of time, the output phase approaches and remains near the input phase, say within 30°. Then the sine function of the input-output phase difference is very nearly the phase difference itself. Equation (9) may then be rewritten, with good approximation, as

$$\varphi_o(t) = \frac{A A_V K'_\varphi K_V}{2} \int_0^t \int_0^\infty h(\tau) \left[\varphi_i(t-\tau) - \varphi_o(t-\tau)\right] d\tau\, dt \tag{10}$$

Equation (10) is linear and easily Laplace transformable. First, assuming the integrand is well behaved, the order of integration may be changed to give

$$\varphi_o(t) = \frac{A A_V K'_\varphi K_V}{2} \int_0^\infty h(\tau) \int_0^t \left[\varphi_i(t-\tau) - \varphi_o(t-\tau)\right] dt\, d\tau \tag{11}$$

Defining the Laplace transform of $\varphi(t)$ to be $\Phi(s)$, equation (11) may be transformed to

$$\Phi_o(s) = \frac{A A_V K'_\varphi K_V}{2} \int_0^\infty h(\tau) \left[\Phi_i(s) - \Phi_o(s)\right] \frac{e^{-s\tau}}{s} d\tau \tag{12}$$

$$\Phi_o(s) = \frac{AA_V K'_\varphi K_V}{2s}\left[\Phi_i(s) - \Phi_o(s)\right]\int_0^\infty h(\tau)e^{-s\tau}d\tau \qquad (13)$$

Now the integral over τ of the loop filter impulse response is simply the Laplace transform or the transfer function of the loop filter which will be labeled $F_L(s)$. Then

$$\Phi_o(s) = \frac{AA_V K'_\varphi K_V}{2}\left[\Phi_i(s) - \Phi_o(s)\right]\frac{F_L(s)}{s} \qquad (14)$$

Next, a gain constant for the loop is defined as

$$K \triangleq \frac{AA_V K'_\varphi K_V}{2} \qquad (15)$$

and equation (13) is rearranged as

$$\frac{\Phi_o(s)}{\Phi_i(s)} = \frac{K\dfrac{F_L(s)}{s}}{1 + K\dfrac{F_L(s)}{s}} \qquad (16)$$

The form of equation (15) is well known from the theory of servomechanisms, representing a linear servo loop whose output $\varphi_o(t)$ is subtracted from its input $\varphi_i(t)$, and having a loop gain function $K\dfrac{F_L(s)}{s}$. The tracking properties of servomechanisms are well known and well documented (ref. 6). Therefore, once the VCO phase of a phase-locked loop is brought sufficiently near the input phase so that the loop operates linearly, then the loop operates as a linear servomechanism for phase. The loop output phase will "track" the input phase within the dynamic capabilities of the servo as determined by the loop filter transfer function $F_L(s)$; hence, the name "phase-locked loop."

C.2 The Linearized Model of the Phase-locked Loop

From the work in the preceding section, it is evident that as long as the output phase tracks the input phase closely enough, say within

30°, the loop may be described for phase by linear transfer functions. Equation C.1 (15), page C-4, may be written in an expanded form from which the equivalent linear model of the loop, for phase, may be drawn by inspection.

$$\frac{\Phi_o(s)}{\Phi_i(s)} = \frac{\left[\frac{AA_V K'_\varphi}{2}\right]\left[\frac{K_V}{s}\right] F_L(s)}{1 + \left[\frac{AA_V K'_\varphi}{2}\right]\left[\frac{K_V}{s}\right] F_L(s)} \quad (1)$$

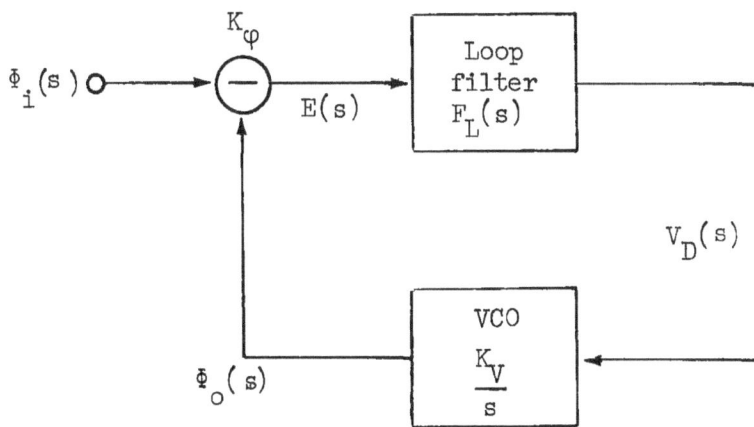

Figure C.2-1.- Linear loop model

Inspection of equation (1) and figure C.2-1 shows that the multiplier has been replaced by a phase subtractor with gain constant K_φ, which is proportional to the multiplier gain constant K'_φ, and the amplitudes of the input signal A and VCO signal A_V.

$$K_\varphi = \frac{AA_V K'_\varphi}{2} \quad (2)$$

C.2.1 The Closed Loop Transfer Functions

From the linear model of the loop three transfer functions of interest may be derived. These three functions relate the input signal phase $\Phi_i(s)$ to the output signal phase $\Phi_o(s)$, the VCO driving signal $V_D(s)$, and the phase error signal $E(s)$.

$$\frac{\Phi_o(s)}{\Phi_i(s)} \triangleq G(s) \equiv \frac{\frac{K_V K_\varphi F_L(s)}{s}}{1 + \frac{K_V K_\varphi F_L(s)}{s}} \qquad (1)$$

$$\frac{E(s)}{\Phi_i(s)} \equiv K_\varphi \left[1 - G(s)\right] \equiv \frac{K_\varphi}{1 + \frac{K_V K_\varphi F_L(s)}{s}} \qquad (2)$$

$$\frac{V_D(s)}{\Phi_i(s)} \equiv \frac{K_\varphi F_L(s)}{1 + \frac{K_V K_\varphi F_L(s)}{s}} \qquad (3)$$

Equations (1), (2), and (3) hold for any loop filter. They may be specialized for the usual loop filter transfer function which has one each, real, finite transmission zero, and pole.

$$F_L(s) \triangleq K_f \frac{(s - z)}{(s - p)} \qquad (4)$$

where

K_f = filter constant, a dimensionless number

The closed loop transfer function may now be written as

$$G(s) \equiv K_V K_\varphi K_f \left[\frac{s - z}{s^2 + s\left(K_V K_\varphi K_f - p\right) - K_V K_\varphi K_f z} \right] \qquad (5)$$

A total gain constant K may now be defined as

$$K \triangleq K_V K_\varphi K_f \qquad (6)$$

The three transfer functions may be rewritten as

$$G(s) = K \left[\frac{s - z}{s^2 + s(K - p) - Kz} \right] \qquad (7)$$

$$\frac{E(s)}{\Phi_i(s)} = K_\varphi \left[\frac{s(s-p)}{s^2 + s(K-p) - Kz} \right] \qquad (8)$$

$$\frac{V_D(s)}{\Phi_i(s)} = K_\varphi K_f \left[\frac{s(s-z)}{s^2 + s(K-p) - Kz} \right] \qquad (9)$$

It is seen that the denominators of the transfer functions are of the form

$$D(s) = s^2 + 2\xi\omega_n s + \omega_n^2 \equiv s^2 + s(K-p) - Kz \qquad (10)$$

where

$$\omega_n = \sqrt{-Kz}$$

$$\xi = \frac{K-p}{2\sqrt{-Kz}} \qquad (11)$$

A valid approximation for most second order loops is that

$$K \gg -p \qquad (12)$$

then

$$\xi \cong \frac{1}{2}\sqrt{-\frac{K}{z}} \qquad (13)$$

and

$$z \cong -\frac{\omega_n}{2\xi} \qquad (14)$$

$$K \cong 2\xi\omega_n \qquad (15)$$

It is informative to draw the asymptotic Bode diagrams of the steady state transfer functions. It should be understood that these diagrams are valid only so long as the loop is locked and is operating linearly. The diagrams apply only to periodic input phase functions. There are certain aperiodic phase functions which will cause the loop to become nonlinear and to unlock. These functions will be examined in a following section.

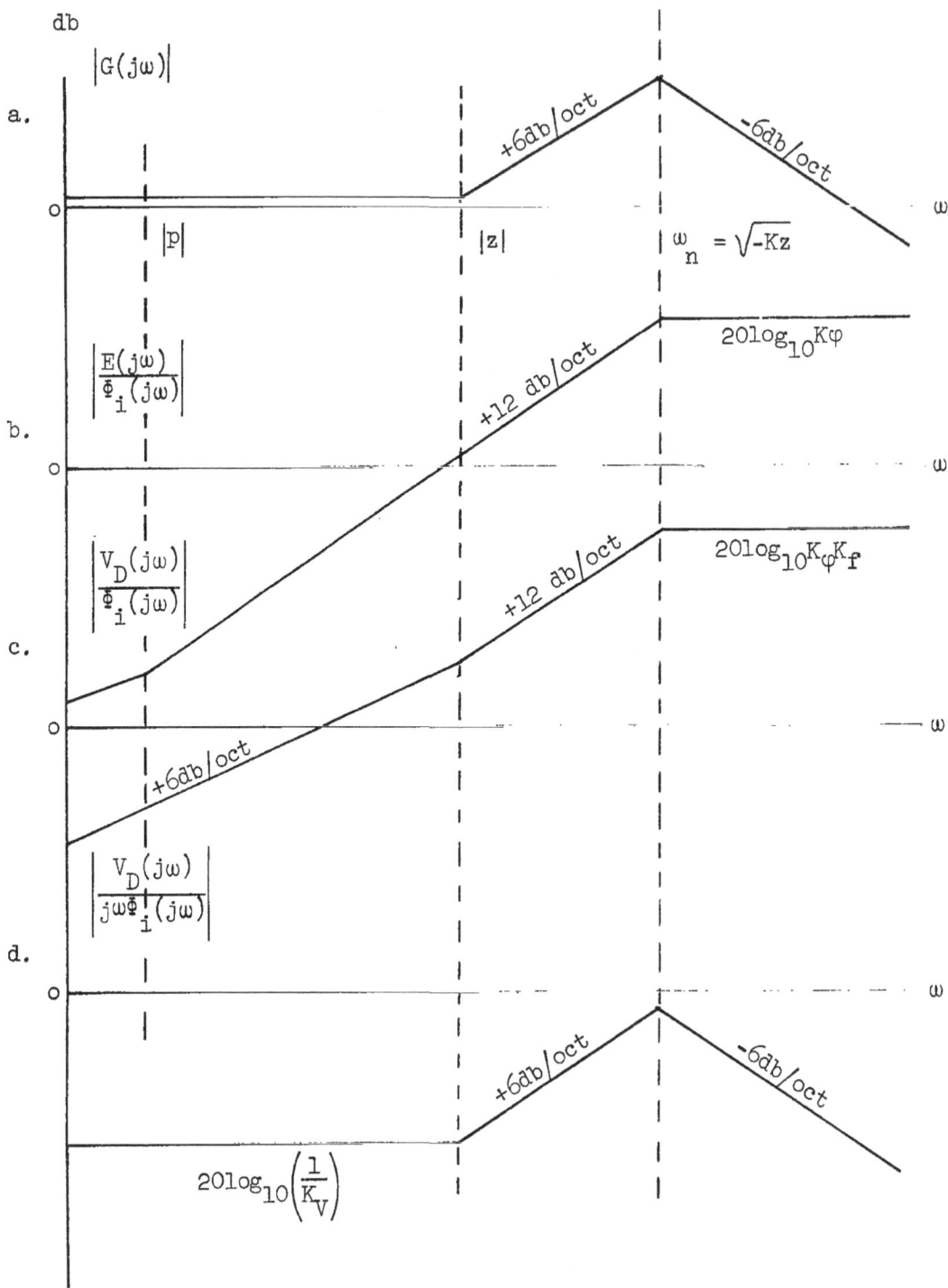

Figure C.2.1-1.- Asymptotic Bode plots of transfer functions

C.2.2 Modulation Tracking Error

It has been stated that the linear treatment of the phase-lock loop is valid only so long as the instantaneous error between input phase $\varphi_i(t)$ and output phase $\varphi_o(t)$ remains small; say less than $30°$. This section examines the effects of certain input signal phase functions on the error function.

$$e(t) = \varphi_i(t) - \varphi_o(t) \tag{1}$$

Although this error analysis is linear, while the very error it attempts to analyze causes the loop to become nonlinear, still, the results are useful for inference of the loop operation. The analysis will be carried through by evaluating the inverse Laplace transform of the error function for each of four input phase signals. The input signals and corresponding Laplace transforms are given in table C.2.2-I.

Case	$\varphi_i(t)$	$\Phi_i(s)$
1	$K_1 U(t)$	$\dfrac{K_1}{s}$
2	$K_2 t U(t)$	$\dfrac{K_2}{s^2}$
3	$K_3 t^2 U(t)$	$2\dfrac{K_3}{s^3}$
4	$U(t) \sum\limits_{n=1}^{K} K_n \sin \omega_n t$	$\sum\limits_{n=1}^{K} \dfrac{K_n \omega_n}{s^2 + \omega_n^2}$

TABLE C.2.2-I INPUT FUNCTIONS

$U(t)$ is the unit step function, as defined as

$$U(t) = 1 \; ; \; t \geq 0$$
$$= 0 \; ; \; t < 0 \tag{2}$$

Case 1 is a phase step input of amplitude K_1 radians. Case 2 is a phase ramp input with slope K_2 radians/second. Case 3 is a phase acceleration with acceleration K_3 radians/second^2. Case 4 is a summation of sine waves having amplitudes K_n and frequencies ω_n.

The error function to be treated is given as

$$E(s) = \Phi_i(s) - \Phi_o(s) \equiv \left[\frac{K_\varphi s(s-p)}{s^2 + s(K-p) - Kz}\right] \Phi_i(s) \qquad (3)$$

Case 1:

$$\varphi_i(t) = K_1 U(t) \qquad (1)$$

$$\Phi_i(s) = \frac{K_1}{s} \qquad (2)$$

$$E(s) = \frac{K_1 K_\varphi (s-p)}{s^2 + s(k-p) - Kz} \qquad (3)$$

$$e(t) = 2K_1 K_\varphi \sqrt{\frac{p-z}{-4z - \frac{1}{K}(K-p)^2}}\, e^{-\left(\frac{K-p}{2}\right)t} \sin\left[\sqrt{-Kz - \frac{(K-p)^2}{4}}\, t + \psi\right] \qquad (4)$$

where

$$\psi = \tan^{-1}\left[\frac{-\sqrt{-4Kz - (K-p)^2}}{K+p}\right] \qquad (5)$$

It is seen that there is no steady-state error, only a transient. The peak transient error may be found by setting the first derivative of $e(t)$ to zero.

Case 2:

$$\varphi_i(t) = K_2 t U(t) \qquad (1)$$

$$\Phi_i(s) = \frac{K_2}{s^2} \qquad (2)$$

$$E(s) = \frac{K_2 K_\varphi (s-p)}{\left[s\, s^2 + s(K-p) - Kz\right]} \qquad (3)$$

$$e(t) = \frac{K_2 K_\varphi}{z} \left\{ \frac{\dot{p}}{K} + 2\sqrt{\frac{p-z}{4K + \frac{(K-p)^2}{z}}} \, e^{-\frac{(K-p)}{2}t} \cdot \sin\left[\sqrt{-Kz - \frac{(K-p)^2}{4}}\, t + \psi\right] \right\} \quad (4)$$

where

$$\psi = \tan^{-1} \sqrt{\frac{-4Kz}{(K-p)^2} - 1} \; - \; \tan^{-1} \frac{2\sqrt{-Kz - \frac{(K-p)^2}{4}}}{K + p} \quad (5)$$

It is seen that there is a transient error and also a steady-state error which is dependent on the pole frequency p of the loop filter. This steady-state error is generally small enough to be neglected.

Case 3:

$$\varphi_i(t) = K_3 t^2 U(t) \quad (1)$$

$$\Phi_i(s) = \frac{2K_3}{s^3} \quad (2)$$

$$E(s) = \frac{2K_3 K_\varphi (s - p)}{s^2 \left[s^2 + s(K-p) - Kz\right]} \quad (3)$$

$$e(t) = \frac{2K_3 K_\varphi}{Kz} \left\{ pt - 1 + \frac{p(K-p)}{Kz} + \sqrt{\frac{K(p-z)}{-Kz - \frac{(K-p)^2}{4}}} \, e^{-\frac{(K-p)}{2}t} \sin\left[\sqrt{-Kz - \frac{(K-p)^2}{4}}\, t + \psi\right] \right\} \quad (4)$$

where

$$\psi = 2\tan^{-1}\left[\frac{-2\sqrt{-Kz - \frac{(K-p)^2}{4}}}{K-p}\right] - \tan^{-1}\left[\frac{-2\sqrt{-Kz - \frac{(K-p)^2}{4}}}{K+p}\right] \quad (5)$$

It is seen that there is a transient error, a steady-state error which is an increasing function of time, dependent on the pole frequency, and two constant steady-state errors, one dependent on the pole frequency.

Case 4:

$$\varphi_i(t) = U(t) \sum_{n=1}^{K} K_n \sin \omega_n t \quad (1)$$

$$\Phi_i(s) = \sum_{n=1}^{K} \frac{K_n \omega_n}{s^2 + \omega_n^2} \quad (2)$$

$$E(s) = \sum_{n=1}^{K} \frac{K_n K_\varphi \omega_n s(s-p)}{\left[s^2 + \omega_n^2\right]\left[s^2 + s(K-p) - Kz\right]} \quad (3)$$

$$e(t) = \sum_{n=1}^{K} K_n K_\varphi \left\{ \frac{-\omega_n \sqrt{\omega_n^2 + p^2}}{\sqrt{\left(\omega_n^2 + Kz\right)^2 + (K-p)^2 \omega_n^2}} \cos\left[\omega_n t + \psi_1\right] \right.$$

$$\left. + \sqrt{\frac{-K(p-z)}{Kz + \frac{(K-p)^2}{4}}} e^{-\frac{(K-p)}{2}t} \cos\left[\sqrt{-Kz - \frac{(K-p)^2}{4}} t + \psi_2\right] \right\}$$

$$(4)$$

where

$$\psi_1 = \tan^{-1}\left[\frac{\omega_n}{-p}\right] - \tan^{-1}\left[\frac{(K-p)\omega_n}{-Kz - \omega_n^2}\right] \quad (5)$$

$$\psi_2 = \tan^{-1}\left[\frac{\sqrt{-Kz - \frac{(K-p)^2}{4}}}{-p - \frac{(K-p)}{2}}\right] - \tan^{-1}\left[\frac{(K-p)\sqrt{-Kz - \frac{(K-p)^2}{4}}}{\frac{1}{2}(K-p)^2 + \omega_n^2 + Kz}\right]$$

$$+ \tan^{-1}\left[\frac{K-p}{2\sqrt{-Kz - \frac{(K-p)^2}{4}}}\right] \quad (6)$$

It is seen that there is a transient error and also a steady-state error which is a sum of sinusoids having frequencies the same as the input sinusoids, and amplitudes and phases which are dependent on sinusoid frequency and loop parameters. The steady-state peak error in radians may be seen to be

$$e_{pk} \; \sum_{i=1}^{K} \frac{\Delta\varphi_i}{\sqrt{1 + \frac{K^2 + 2Kz}{\omega_i^2} + \frac{(Kz)^2}{\omega_i^2}}} \quad (7)$$

$$e_{pk} \cong \sum_{i=1}^{K} \frac{\Delta\varphi_i}{\sqrt{1 + \frac{2\omega_n^2(2\xi^2 - 1)}{\omega_i^2} + \frac{\omega_n^4}{\omega_i^4}}} \quad (8)$$

C.2.3 Loop Phase Noise

In the following sections it will be necessary to be able to relate the phase noise or "jitter" of the VCO signal to the noise accompanying the input signal. If the input phase noise is characterized as being Gaussian and has a flat spectral density with value $|\Phi_\varphi|$, then the variance σ_φ^2 or mean squared value of the VCO phase noise may be written as the product of the input spectral density times the "closed-loop equivalent noise bandwidth" B_N. For two-sided spectral densities we may write

$$\sigma_\varphi^2 = \left|\Phi_\varphi\right| 2B_N \tag{1}$$

The closed-loop equivalent noise bandwidth may be computed in terms of general loop parameters, using the loop transfer function of equation C.2.1 (7), page C-6, and the method of appendix H-1.

$$\frac{\Phi_o(s)}{\Phi_i(s)} = G(s) = \frac{K(s - z)}{s^2 + s(K - p) - Kz} \tag{2}$$

Let

$$G(s) = \frac{K(s - z)}{(S - A)(s - B)} \tag{3}$$

where

$$A = -\frac{(K - p)}{2} + \frac{1}{2}\sqrt{(K - p)^2 + 4Kz} \triangleq -\alpha + \beta \tag{4}$$

$$B = -\frac{(K - p)}{2} - \frac{1}{2}\sqrt{(K - p)^2 + 4Kz} \triangleq -\alpha - \beta \tag{5}$$

Then,

$$G(s)G(-s) = \left[\frac{K(s - z)}{(s - A)(s - B)}\right]\left[\frac{-K(s + z)}{(s + A)(s + B)}\right] \tag{6}$$

It is seen that the only poles in the left half plane are at $s = A$ and $s = B$, respectively. Define

$$G(s)G(-s) \triangleq \psi(s) \tag{7}$$

The residue at $s = A$ is given as

$$\operatorname{Res} \psi(s)\Big|_{s=A} = \lim_{s \to A} (s - A)\,\psi(s) = \frac{-K^2(A^2 - z^2)}{(A - B)(2A)(A + B)} \tag{8}$$

The residue at $s = B$ is given as

$$\operatorname{Res} \psi(s)\Big|_{s=B} = \lim_{s \to B} (s - B)\,\psi(s) = \frac{-K^2(B^2 - z^2)}{(B - A)(2B)(B + A)} \tag{9}$$

The sum of the residues in the left half plane is given as

$$\sum \text{Res } \psi(s) \bigg|_{\text{LHP}} = \frac{\frac{K^2}{2B}\left(B^2 - z^2\right) - \frac{K^2}{2A}\left(A^2 - z^2\right)}{A^2 - B^2} \tag{10}$$

$$\sum \text{Res } \psi(s) \bigg|_{\text{LHP}} = \frac{K^2}{4} \frac{\alpha^2 - \beta^2 + z^2}{\alpha\left(\alpha^2 - \beta^2\right)} \tag{11}$$

Substituting and reducing, we have

$$\frac{1}{G_{\text{ref}}^2} \sum \text{Res } \psi(s) \bigg|_{\text{LHP}} = \frac{1}{G_{\text{ref}}^2} \frac{K}{2} \left[\frac{K - z}{K - p}\right] \tag{12}$$

For this loop the low frequency gain is taken as reference

$$G_{\text{ref}} = \lim_{s \to 0} G(s) = 1 \tag{13}$$

Then

$$2\Delta\omega_N = \pi K \left[\frac{K - z}{K - p}\right] \tag{14}$$

or

$$2B_N = \frac{K}{2} \left[\frac{K - z}{K - p}\right] \tag{15}$$

For the alternate notation,

$$K = 2\xi\omega_n \; ; \; z = -\frac{\omega_n}{2\xi} \tag{16}$$

$$2B_N = \frac{\omega_n}{4\xi} \left[1 + 4\xi^2\right] \tag{17}$$

C.2.4 Threshold Prediction

A phase-lock loop is useful only when it is locked. A phase-locked loop which is operating at a high input signal-to-noise ratio will remain locked most of the time. As the input signal-to-noise ratio is lowered the loop will break lock more frequently, but will regain lock if the signal-to-noise ratio is not too low.

Perhaps the simplest way to treat threshold is to define the loop as operating above threshold if it is in lock a certain average percent of the time and define it as below threshold if it is in lock less than the required percent of time. In this manner loop threshold is rather subjective and is dependent on the loop's use, which defines the threshold in lock time percentage.

The analytical methods and assumptions by which the signal, noise, and loop parameters are related to the percent inlock time have provided a fertile field for analysis. Martin (ref. 7) defined a "practical" "absolute" threshold, predictable from a linear loop model, which was reasonably substantiated by laboratory test data. Later work by Develet (ref. 8) treated threshold with nonlinear loop models.

It is the purpose of this section to set down the simplest method of loop treatment which will yield results of tolerable accuracy. The simplest method is to define the conditions under which the linear loop model is valid and then use the linear model to infer the nonlinear threshold properties of the loop.

An assumption which highly simplifies the analysis is that the phase component of the input signal is separable into a distinct signal term and a distinct noise term. A second simplifying assumption is that the phase noise term represents a Gaussian noise process having a flat spectral density. With these two assumptions the modulation tracking error of the loop in response to the input signal phase term and the VCO phase jitter in response to the input phase noise term may be easily determined by the methods set forth in sections C.2.2 and C.2.3.

With a knowledge of the modulation tracking error, and especially the peak tracking error e_m in radians, and a knowledge of the standard deviation σ_φ of the VCO phase noise in radians, Martin's (ref. 7) threshold criterion may be employed. It is given an

$$e_m + x\sigma_\varphi \leq \frac{\pi}{2} \qquad (1)$$

where x is a peak factor or confidence factor for the VCO phase noise.

The significance of the number $\frac{\pi}{2}$ may be seen by observation of equation C.1 (9), page C-3, the nonlinear equation giving the loop response to input phase. Suppose initially both $\varphi_i(t)$ and $\varphi_o(t)$ are identically zero. Suppose $\varphi_i(t)$ increases positively from zero. Then, according to equation (9) $\varphi_o(t)$ will increase positively to track $\varphi_i(t)$. This tracking is caused by the error function $\sin\left[\varphi_i(t) - \varphi_o(t)\right]$, increasing as $\varphi_i(t)$ separates in value from $\varphi_o(t)$. However, if $\varphi_i(t)$ separates from $\varphi_o(t)$ rapidly enough so that the instantaneous value of $\varphi_i(t) - \varphi_o(t)$ exceeds $\frac{\pi}{2}$ radians, then the error function $\sin[\varphi_i(t) - \varphi_o(t)]$ will decrease with increasing phase error and $\varphi_o(t)$ will not track $\varphi_i(t)$. In other words, if the instantaneous value of $\varphi_i(t) - \varphi_o(t)$ exceeds $\frac{\pi}{2}$ for a loop which is initially locked, the loop will break lock.

Given a peak tracking error e_m, then the statistical probabability of the loop breaking lock is implied by equation (1) above. For Gaussian phase jitter the probability of the loop breaking lock at the time of peak tracking error e_m may be determined. In probabilistic notation, the probability that the locked loop loses lock is given by

$$P\left[x\pi_\varphi + e_m > \frac{\pi}{2}\right] = 1 - P\left[x\sigma_\varphi + e_m < \frac{\pi}{2}\right] \qquad (2)$$

$$P\left[x\sigma_\varphi + e_m > \frac{\pi}{2}\right] = 1 - \Phi_x(x) \; ; \; x = \frac{\frac{\pi}{2} - e_m}{\sigma_\varphi} \qquad (3)$$

where

$\Phi_x(x)$ = normal distribution function.

In terms of the error function, which is tabulated, we have

$$P\left[x\sigma_\varphi + e_m > \frac{\pi}{2}\right] = \frac{1}{2}\left[1 - \text{erf}\left(\frac{x}{\sqrt{2}}\right)\right] \qquad (4)$$

Equations (3) and (4) show that given a tracking error e_m, the standard deviation σ_φ of the VCO phase jitter uniquely determines the probability of loss of lock. Given a required probability of loss of lock, the confidence value x may be obtained from tables of the error function.

Table C.2.4-I gives the values of x corresponding to loss of lock probabilities for five cases.

P	10^{-1}	10^{-2}	10^{-3}	10^{-4}	10^{-5}
x	1.29	2.34	3.1	3.72	4.25

TABLE C.2.4-I.- CONFIDENCE VALUES VERSUS LOSS-LOCK PROBABILITIES

In sections C.4 and C.5 to follow, detailed threshold relations will be derived for two special cases of phase-locked loops, relating the probability of loss of lock to the input signal-to-noise ratio.

C.3 Signal and Noise Characteristics of Prelimited Phase-Locked Loops

The remainder of appendix C will be restricted to phase-locked loops preceeded by an ideal band-pass limiter. The model is given in figure C.3-1.

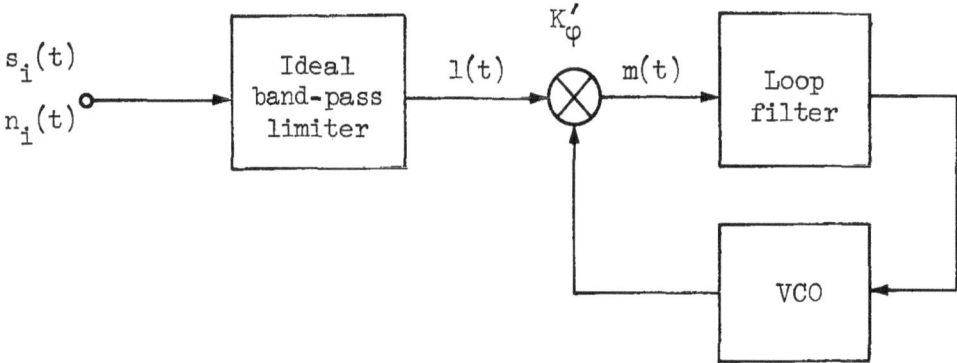

Figure C.3-1.- Prelimited phase-locked loop

The input signal $s_i(t)$ is taken as an ideal angle modulated signal from equation A.1 (7), page A-2.

$$s_i(t) = A \cos \left[\omega_c t + \varphi_s(t) \right] \tag{1}$$

The noise is taken in the form of equation B.1 (1), page B-1, as a sample function of a Gaussian process, with a flat spectral density band-limited to the input bandwidth of the limiter.

$$n_i(t) = x(t) \cos \omega_c t - y(t) \sin \omega_c t \tag{2}$$

The limiter output function is taken from equation B.3 (4), page B-6, as

$$l(t) = 4 \frac{V_L}{\pi} \cos \left[\omega_c t + \psi(t) \right] \tag{3}$$

where, from equation B.2 (9), page B-4,

$$\psi(t) = \arctan\left[\frac{A \sin \varphi_s(t) + y(t)}{A \cos \varphi_s(t) + x(t)}\right] \tag{4}$$

For high signal-to-noise ratio (SNR) into the limiter the output function is approximated, from equation B.3 (5), page B-6, as

$$l(t) = 4\frac{V_L}{\pi} \cos\left[\omega_c t + \varphi_s(t) + \frac{y(t)}{A}\right] \tag{5}$$

For high input SNR into the limiter, it is seen from equation (5) that the input signal phase function to the phase-locked loop is $\varphi_s(t)$.

Also, from equation B.2 (11), page B-5, the phase noise spectral density at the input to the loop is

$$\Phi_\varphi(\omega) = \frac{2\left|\Phi_{n_i}\right|}{A^2} \; ; \; |\omega| < \frac{\Delta\omega}{2} \tag{6}$$

where

$$\Delta\omega = 2\pi B_i \tag{7}$$

$\left|\Phi_{n_i}\right|$ is the constant value of the flat noise spectral density into the limiter and B_i is the limiter input bandwidth.

C.3.1 Limiter Effects on Loop Parameters

Section H.3, page H-9, which is based on Davenport's (ref. 4) work, discloses a property of band-pass limiters which affects the parameters of a limiter driven phase-locked loop. At low limiter SNR the amplitude of the sinusoid feeding the phase-locked loop is suppressed by a factor α_L from its value at high limiter SNR. Martin's (ref. 7) approximation to α_L is reproduced here from equation H.3(3), page H-11

$$\alpha_L^2 = \frac{1}{1 + \frac{4}{\pi}\left[\frac{N_i}{S_i}\right]} \tag{1}$$

where

$\left[\frac{N_i}{S_i}\right]$ = limiter input noise-to-signal ratio

Equation C.2 (2), page C-5, shows that the phase subtractor constant K_φ is proportional to the amplitude of the sinusoid feeding the phase-locked loop. When the limiter suppresses the sinusoid, it also suppresses K_φ by the same factor α_L. The value of K_φ which is suppressed by the limiter action will be denoted by K_{φ_α}.

then

$$K_{\varphi_\alpha} = \alpha_L K_\varphi \tag{2}$$

The loop parameters derived in section C.2.1 and C.2.3 may be modified for limiter suppression. Since the phase subtractor gain is reduced, so is the loop gain K. If K is understood to be the maximum loop gain for no limiter suppression, then the modified gain is

$$K_\alpha = \alpha_L K \tag{3}$$

and

$$\varepsilon_\alpha \cong \frac{1}{2}\sqrt{-\frac{\alpha_L K}{z}} = \xi\sqrt{\alpha_L} \tag{4}$$

$$\omega_{n_\alpha} = \sqrt{-\alpha_L K z} = \omega_n \sqrt{\alpha_L} \tag{5}$$

$$2B_{N_\alpha} = \frac{\alpha_L K}{2}\left[\frac{\alpha_L K - z}{\alpha_L K - p}\right] \tag{6}$$

It should be noted that the tracking error of the loop changes in a like manner. The errors for particular modulations and particular alphas may be evaluated through use of the expressions of section C.2.2.

C.4 Modulation Restrictive Loop

This section will consider the threshold treatment of a special type of loop known as "modulation restrictive". This type loop is used to track an unmodulated sinusoidal carrier or the residual carrier component of a narrow-phase modulated signal. It is assumed that by suitable input filtering and proper signal design, a modulation restrictive loop will see little or no signal modulation and will operate with negligible modulation error e_m except for that error caused by Doppler effect.

The following treatment will be for the special case of zero Doppler effect. If Doppler effect cannot be neglected, loop threshold may be treated easily using the results of section C.2.4.

Several assumptions are made. First, the limiter bandwidth is taken to be much wider than the loop equivalent noise bandwidth. Second, the statistics of the phase noise process passing from the limiter into the loop are assumed to be approximately Gaussian for any limiter SNR. The effect of limiter suppression is included in the determination of the closed loop noise bandwidth B_N.

For this special case, the threshold defining equation C.2.4 (1), page C-16, specializes to

$$x\sigma_\varphi = \frac{\pi}{2} \qquad (1)$$

where the equality defines the threshold VCO phase jitter. Dividing equation (1) by x and squaring,

$$\sigma_\varphi^2 = \left[\frac{\frac{\pi}{2}}{x}\right]^2 \qquad (2)$$

From equation C.2.3 (1), page C-14,

$$\sigma_\varphi^2 = \left|\Phi_\varphi\right| 2B_N \qquad (3)$$

From equation C.3 (6), page C-19,

$$\left|\Phi_\varphi\right| = \frac{2\left|\Phi_{ni}\right|}{A^2} \qquad (4)$$

and

$$\sigma_\varphi^2 = \frac{\left|\Phi_{ni}\right| 2B_N}{\frac{A^2}{2}} \qquad (5)$$

But the quantity on the right-hand side of equation (5) is identically the noise-to-signal ratio into the limiter, computed in the loop noise bandwidth B_N.

then

$$\sigma_\varphi^2 = \left[\frac{N_i}{S_i}\right]_{B_N} \qquad (6)$$

and

$$\left[\frac{S_i}{N_i}\right]_{B_N} = \left[\frac{x}{\frac{\pi}{2}}\right]^2 \qquad (7)$$

Thus, the limiter input SNR, taken in the loop noise bandwidth, has been related to the confidence value x. From the results of table C.2.4-I, page C-17, the required SNR may be tabulated for various probabilities of loss of phase-lock. These are as follows:

P	10^{-1}	10^{-2}	10^{-3}	10^{-4}	10^{-5}
$\left[\frac{S_i}{N_i}\right]_{B_N}$	-1.7db	3.46db	5.91db	7.47db	8.62db

TABLE C.4-I.- INPUT SNR VERSUS LOSS-LOCK PROBABILITIES

It should be noted that when making computations involving B_N, the value of B_N used should be that value actually produced by the limiter input SNR. The calculation of B_N is treated in the following section.

C.4.1 Loop Noise Bandwidth Above Threshold

For a prelimited modulation restrictive phase-locked loop which has been optimized (for threshold) at some particular loop signal-to-noise ratio, it is necessary to be able to determine the effective loop noise bandwidth for signal-to-noise ratios above threshold. The usual assumptions are made that the loop is locked, fed by an ideal limiter, an unmodulated sinusoid, and white band-limited Gaussian noise. From equation C.2.3 (15), page C-15, the two sided closed loop noise bandwidth is taken as

$$2B_N = \frac{K}{2}\left[\frac{K - z}{K - p}\right] \qquad (1)$$

where

K = open loop gain

z = loop filter zero frequency

p = loop filter pole frequency

Results will be obtained for a special case which represents a wide class of loops. The loop filter parameters will be set such that

$$|p| \ll K$$
$$\xi = 0.707 \qquad (2)$$

where

ξ = loop damping factor

then

$$z \triangleq z_o \equiv -\frac{K_o}{2}$$

$$K_o = \alpha_o K_H \qquad (3)$$

where

α = limiter signal voltage suppression factor

K_H = maximum or high signal value of loop gain

$_o$ = conditions at the loop design threshold

Then

$$2B_N = \frac{1}{2}\left[K + \frac{K_o}{2}\right] = \frac{K_H}{2}\left[\alpha + \frac{\alpha_o}{2}\right] \qquad (4)$$

Now the threshold loop noise bandwidth occurs for $\alpha \equiv \alpha_o$, so that

$$2B_{N_o} = \frac{3}{4} \alpha_o K_H \qquad (5)$$

The ratio of bandwidth for any α to threshold bandwidth is

$$\frac{2B_N}{2B_{N_o}} = \frac{1}{3}\left[2\frac{\alpha}{\alpha_o} + 1\right] \qquad (6)$$

This result is given by Martin (ref. 7).

Using Martin's (ref. 7) approximation to α as a function of limiter input signal-to-noise ratio, we have

$$\alpha \cong \frac{1}{\sqrt{1 + \frac{4}{\pi} \frac{N}{S}\bigg|_{B_i}}} \qquad (7)$$

or

$$\frac{\alpha}{\alpha_o} \cong \sqrt{\frac{1 + \frac{4}{\pi} \frac{N}{S}\bigg|_{B_{i_o}}}{1 + \frac{4}{\pi} \frac{N}{S}\bigg|_{B_i}}} \qquad (8)$$

then

$$\frac{2B_N}{2B_{N_o}} = \frac{1}{3}\left[2\sqrt{\frac{1 + \frac{4}{\pi} \frac{N_i}{S_i}\bigg|_{B_{i_o}}}{1 + \frac{4}{\pi} \frac{N_i}{S_i}\bigg|_{B_i}}} + 1\right] \qquad (9)$$

A related problem is that of determining the loop signal-to-noise ratio at which a loop was optimized, given a plot of loop bandwidth versus input signal. It is easily determinable that

$$\frac{S_i}{N_i}\bigg|_{B_{i_o}} = \frac{\frac{4}{\pi}}{\left[\frac{3}{2}\frac{B_{N_H}}{B_{N_o}} - \frac{1}{2}\right]^2 - 1} \qquad (10)$$

where

B_{N_H} = strong signal value of loop noise bandwidth

C.5 Prefiltered Modulation Tracking Loops

This section will consider the threshold treatment of a special type of loop known as "prefiltered modulation tracking." This type of loop is used to demodulate angle modulated sinusoidal carriers. For some types of signal modulations it is possible to reduce peak modulation tracking error e_m to a very small value by making the loop natural resonant frequency ω_n very large compared to the highest modulating frequency. Simultaneously, to reduce the transmission of phase noise or jitter through the loop, the input signal plus noise may be processed through a sharp cut-off band-pass filter having a bandwidth just wide enough to pass the modulated signal. For high input signal-to-noise ratios and linear loop operation, the bandwidth of the equivalent phase noise will be half the input bandwidth. For modulation indices not too large, the bandwidth of the phase noise will form, essentially, the closed loop noise bandwidth. Such a modulation tracking loop is called a prefiltered loop.

Assuming the loop natural frequency ω_n is much larger than the highest modulation frequency, the peak modulation tracking error becomes negligibly small. The threshold defining equation for the prefiltered modulation tracking loop then becomes the same as for the modulation restrictive loop, since again e_m approaches zero.

$$x\sigma_\varphi = \frac{\pi}{2} \tag{1}$$

then

$$\sigma_\varphi^2 = \left[\frac{\frac{\pi}{2}}{x}\right]^2 \tag{2}$$

Due to the prefiltering and the assumption that the input bandwidth is much less than the loop ω_n, the effective noise bandwidth of the phase noise is half the input bandwidth or $\frac{B}{2}$. Then, from equation C.2.3 (1), page C-14,

$$\sigma_\varphi^2 = |\Phi_\varphi| 2 \frac{B_i}{2} \tag{3}$$

From equation C.3 (6), page C-19,

$$|\Phi_\varphi| = \frac{2|\Phi_{n_i}|}{A^2} \tag{4}$$

and

$$\sigma_\varphi^2 = \frac{|\Phi_{n_i}| 2 \frac{B_i}{2}}{A^2} \tag{5}$$

then

$$\sigma_\varphi^2 = \frac{1}{2}\left[\frac{N_i}{S_i}\right]_{B_i} \qquad (6)$$

where

$\left[\dfrac{N_i}{S_i}\right]_{B_i}$ = noise-to-signal ratio into the limiter, computed in the limiter bandwidth B_i

then

$$\left[\frac{S_i}{N_i}\right]_{B_i} = \frac{1}{2}\left[\frac{x}{\frac{\pi}{2}}\right]^2 \qquad (7)$$

From the results of table C.2.4-I, page C-19, the required SNR into the limiter may be tabulated for various probabilities of loss of phase-lock. These are given as follows:

P	10^{-1}	10^{-2}	10^{-3}	10^{-4}	10^{-5}
$\dfrac{S_i}{N_i}_{B_i}$	-4.7db	0.46db	2.91db	4.47db	5.62db

TABLE C.5-I.- INPUT SNR VERSUS LOSS-LOCK PROBABILITIES

It is seen that for the lower probabilities the limiter input SNR is low enough to violate the assumption of high SNR which was used to obtain the phase noise spectral density. Therefore, care should be exercised in applying the results of table I for the lower SNR.

APPENDIX D

PRODUCT DEMODULATION

D.1 Linear Product Demodulator

Figure D.1-1 shows the configuration of a product demodulator, used to coherently detect phase modulation. This is the product detector of section B.4.1, page B-6, followed by an ideal output filter.

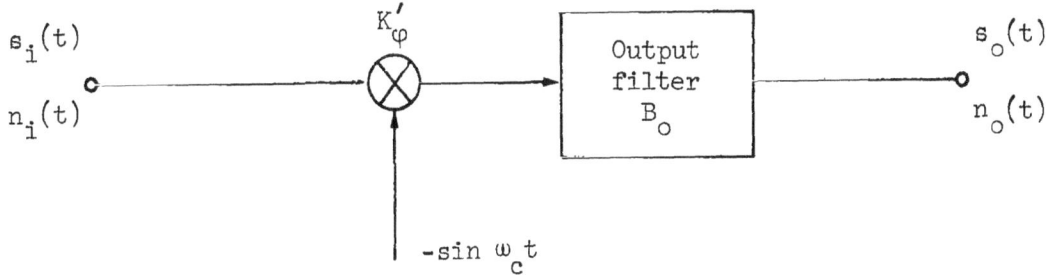

Figure D.1-1.- Demodulator configuration

The output filter is defined to be either low-pass or band-pass, with flat unity amplitude and phase transmission characteristics, square frequency cut-off characteristics, and transmission bandwidth of B_o cps.

The input signal and noise are taken in the usual forms as

$$s_i(t) = A \cos \left[\omega_c t + \varphi_s(t) \right] \qquad (1)$$

$$n_i(t) = x(t) \cos \omega_c t - y(t) \sin \omega_c t \qquad (2)$$

where

$s_i(t)$ = angle modulated signal with equivalent phase modulation $\varphi_s(t)$

$n_i(t)$ = sample function of a random Gaussian process

$n_i(t)$ is further defined as white and band-limited to B cps.

The noise power spectral density at the multiplier output is given from equation B.4.1 (7), page B-7, as

$$\Phi_m(\omega) = \frac{K_\varphi'^2}{2} \left|\Phi_{n_i}\right| \; ; \; |\omega| < 2\pi \frac{B}{2}$$

$$= 0 \quad ; \text{ all other values of } \omega \qquad (3)$$

where

$$\left|\Phi_{n_i}\right| = \text{constant value of the input noise spectral density } \Phi_{n_i}(\omega)$$

The signal component, from equation B.4.1 (4), page B-7, is

$$s_m(t) = \frac{K_\varphi'}{2} A \sin \varphi_s(t) \qquad (4)$$

Equation (4) will be subsequently treated for specific signal types.

D.1.1 Detection of Sinusoidal Subcarriers

This section treats demodulation of a carrier which is phase modulated by the sum of a pseudo-random range code plus K subcarriers. The model of figure D.1-1 applies.

The input signal is taken in usual form as

$$s_i(t) = A \cos \left[\omega_c t + \Delta\varphi_r c_t(t) + \sum_{i=1}^{K} \Delta\varphi_i \sin\left(\omega_i t + \varphi_i\right) \right] \qquad (1)$$

the multiplier output is

$$m(t) \cong K_\varphi' \frac{A^2}{2} \sin \left[\Delta\varphi_r c_t(t) + \sum_{i=1}^{K} \Delta\varphi_i \sin\left(\omega_i t + \varphi_i\right) \right] \qquad (2)$$

Expanding equation (2) and applying the identities of equations A.3 (3) and A.3 (4), page A-7, we obtain

$$m(t) \cong K'_\varphi \frac{A}{2} \left\{ c_t(t) \sin\left(\Delta\varphi_r\right) \cos\left[\sum_{i=1}^{K} \Delta\varphi_i \sin\left(\omega_i t + \varphi_i\right)\right] \right.$$

$$\left. + \cos\left(\Delta\varphi_r\right) \sin\left[\sum_{i=1}^{K} \Delta\varphi_i \sin\left(\omega_i t + \varphi_i\right)\right] \right\} \qquad (3)$$

The first term of equation (3) is a function of the subcarriers multiplied by the range code and interferes with the desired signal which is the second term of equation (3).

Then, the desired multiplier term is

$$m(t) \cong K'_\varphi \frac{A}{2} \cos\left(\Delta\varphi_r\right) \sin\left[\sum_{i=1}^{K} \Delta\varphi_i \sin\left(\omega_i t + \varphi_i\right)\right] \qquad (4)$$

Using the Giacoletto expansion (ref. 1)

$$m(t) \cong K'_\varphi \frac{A}{2} \cos\left(\Delta\varphi_r\right) \left\{ \sum_{n_1=-\infty}^{\infty} \cdots \sum_{n_k=-\infty}^{\infty} \prod_{i=1}^{K} \left[J_{n_i}\left(\Delta\varphi_i\right)\right] \cdot \right.$$

$$\left. \sin\left[\sum_{i=1}^{K} n_i\left(\omega_i t + \varphi_i\right)\right] \right\}$$

$$(5)$$

From equation (5) the jth detected subcarrier terms, the first order terms having frequency ω_j, may be obtained by setting

$$n_i = 0 \quad ; \ i \neq j$$

$$n_i = n_j = \pm 1 \ ; \ i = j \qquad (6)$$

then, for $n_j = +1$

$$m(t)_{n_j=+1} = \frac{K'_\varphi A}{2} \cos(\Delta\varphi_r) J_1(\Delta\varphi_j) \prod_{\substack{i=1 \\ i \neq j}}^{K} \left[J_0(\Delta\varphi_i) \right] \sin(\omega_j t + \varphi_j) \qquad (7)$$

For $n_j = -1$

$$m(t)_{n_j=-1} = \frac{K'_\varphi A}{2} \cos(\Delta\varphi_r) J_{-1}(\Delta\varphi_j) \prod_{\substack{i=1 \\ i \neq j}}^{K} \left[J_0(\Delta\varphi_i) \right] \sin(-\omega_j t - \varphi_j) \qquad (8)$$

Now

$$m_j(t) = m(t)_{n_j=+1} + m(t)_{n_j=-1} \qquad (9)$$

and

$$J_{-n}(x) = (-1)^n J_n(x) \qquad (10)$$

then

$$m_j(t) = K'_\varphi A \cos(\Delta\varphi_r) J_1(\Delta\varphi_j) \prod_{\substack{i=1 \\ i \neq j}}^{K} \left[J_0(\Delta\varphi_i) \right] \sin(\omega_j t + \varphi_j) \qquad (11)$$

As in appendix A, equation (11) holds whether or not the individual subcarriers are, themselves, angle modulated.

The output signal from the output filter is

$$s_{oj}(t) = m_j(t) = K'_\varphi A \cos(\Delta\varphi_r) J_1(\Delta\varphi_j) \prod_{\substack{i=1 \\ i \neq j}}^{K} \left[J_0(\Delta\varphi_i) \right] \sin(\omega_j t + \varphi_j) \qquad (12)$$

D.1.2 Detection of Arbitrary Baseband Modulation

This section treats baseband modulation which is not a sinusoid or sum of sinusoids. The most workable method is to describe the arbitrary function by its peak phase deviation and by an empirically determined peak to rms ratio, or form factor.

The model of figure D.1-1 applies. The input signal is taken as

$$s_i(t) = A \cos \left[\omega_c t + \varphi_s(t)\right] \tag{1}$$

where now $\varphi_s(t)$ is a baseband signal, having peak phase deviation $\Delta\varphi_\rho$.

The output signal from the output filter is

$$s_o(t) = \frac{K'_\varphi}{2} A \sin \varphi_s(t) \tag{2}$$

The peak squared output signal is

$$\left[s_{o\ peak}\right]^2 = \frac{K'^2_\varphi A^2}{4} \sin^2 \left(\Delta\varphi_\rho\right) \tag{3}$$

It is noted that for the output signal to be a linear replica of the phase modulation, $\Delta\varphi_\rho$ must be less than about 30°. If linearity is not of great consideration, as for clipped speech, $\Delta\varphi_\rho$ may be increased. In no case may $\Delta\varphi_\rho$ be greater than 90°. Residual carrier suppression considerations will generally limit $\Delta\varphi_\rho$ to less than 90°.

It is desirable to place a bound on the residual carrier remaining after modulation since generally the demodulator reference signal is derived from the residual carrier. For square wave modulation the remaining carrier is given by the limiting case of equation A.3 (7), page A-8, for $\Delta\varphi_i$ identically zero, as

$$s_c(t) = A \cos \left(\Delta\varphi_\rho\right) \cos \omega_c t \tag{4}$$

Likewise, for sinusoidal modulation, where $\Delta\varphi_r$ is identically zero, the residual carrier is

$$s_c(t) = A\ J_o \left(\Delta\varphi_\rho\right) \cos \omega_c t \tag{5}$$

For the sake of simple analysis, it is assumed that for arbitrary baseband narrow deviation modulation of peak deviation $\Delta\varphi_\rho$, the residual carrier term is bounded by equations (4) and (5).

D.1.3 Noise Characteristics

The noise spectrum out of the multiplier is flat within the limits set by the input bandwidth B. Therefore, the bandwidths B_o of the low-pass or band-pass filters are the equivalent noise bandwidths at the output.

The output noise powers for both the band-pass and low-pass cases are given by

$$N_o = |\Phi_m(\omega)| \, 2B_o \tag{1}$$

$$N_o = \frac{{K'_\varphi}^2}{2} |\Phi_{n_i}| \, 2B_o \tag{2}$$

where

$|\Phi_{n_i}|$ = flat amplitude of the input white noise spectrum

D.1.4 Output Signal-to-noise Ratios

The results of the prior three sections may now be integrated to give output signal-to-noise ratios for coherent demodulation of both subcarriers and arbitrary baseband modulation.

D.1.4.1 Subcarrier and band-pass filter.—
The signal-to-noise ratio out of the ideal band-pass filter B_o cycles wide, for the jth subcarrier signal may now be determined.

From equation D.1.1 (12), page D-4, the output signal for the jth subcarrier is

$$s_{o_j}(t) = K'_\varphi A \cos(\Delta\varphi_r) \, J_1(\Delta\varphi_j) \prod_{\substack{i=1 \\ i \neq j}}^{K} J_o(\Delta\varphi_i) \, \sin(\omega_j t + \varphi_j) \tag{1}$$

The output signal power is then

$$S_{o_j} = {K'_\varphi}^2 \frac{A^2}{2} \cos^2(\Delta\varphi_r) \, J_1^2(\Delta\varphi_j) \prod_{\substack{i=1 \\ i \neq j}}^{K} J_o^2(\Delta\varphi_i) \tag{2}$$

From equation D.1.3 (2), page D-6, the output noise power is

$$N_o = \frac{K_\varphi'^2}{2} \left|\Phi_{n_i}\right| 2B_o \tag{3}$$

The output signal-to-noise ratio is then

$$\frac{S_{o_j}}{N_o} = \frac{K_\varphi'^2 \frac{A^2}{2} \cos^2(\Delta\varphi_r) J_1^2(\Delta\varphi_j) \prod_{\substack{i=1 \\ i\neq j}}^{K} J_o^2(\Delta\varphi_i)}{K_\varphi'^2 \left|\Phi_{n_i}\right| 2B_o} \tag{4}$$

$$\frac{S_{o_j}}{N_o} = 2\cos^2(\Delta\varphi_r) J_1^2(\Delta\varphi_j) \prod_{\substack{i=1 \\ i\neq j}}^{K} J_o^2(\Delta\varphi_i) \left[\frac{\frac{A^2}{2}}{\left|\Phi_{n_i}\right| 2B_o}\right] \tag{5}$$

The quantity in brackets is seen to be the total input signal-to-noise ratio computed in a physical bandwidth B_o. Then

$$\frac{S_{o_j}}{N_o} = 2\cos^2(\Delta\varphi_r) J_1^2(\Delta\varphi_j) \prod_{\substack{i=1 \\ i\neq j}}^{K} \left[J_o^2(\Delta\varphi_i)\right] \left[\frac{S_i}{N_i}\right]_{B_o} \tag{6}$$

As in appendix A, equation (6) holds, whether or not the subcarriers are, themselves, angle modulated.

D.1.4.2 <u>Baseband modulation and low-pass filter.</u>- Proceeding as in the previous section, the peak squared output signal is taken from equation D.1.2 (3), page D-5, as

$$\left[S_{o\ peak}\right]^2 = \frac{K_\varphi'^2 A^2}{4} \sin^2(\Delta\varphi_\rho) \tag{1}$$

Again, from equation D.1.3 (2), page D-6, the output noise power is

$$N_o = \frac{K_\varphi'^2}{2} \left| \Phi_{n_i} \right| 2B_o \tag{2}$$

The peak squared signal to mean-squared noise ratio is then

$$\frac{\left[s_{o\ peak}\right]^2}{N_o} = \frac{\frac{K_\varphi'^2}{2} A^2 \sin^2\left(\Delta\varphi_p\right)}{\frac{K_\varphi'^2}{2} \left|\Phi_{n_i}\right| 2B_o} \tag{3}$$

$$\frac{\left[s_{o\ peak}\right]^2}{N_o} = \sin^2\left(\Delta\varphi_p\right) \left[\frac{\frac{A^2}{2}}{\left|\Phi_{n_i}\right| 2B_o}\right] \tag{4}$$

The bracketed quantity is seen to be the total input signal-to-noise ratio computed in a physical bandwidth B_o. Then

$$\frac{\left[s_{o\ peak}\right]^2}{N_o} = \sin^2\left(\Delta\varphi_p\right) \left[\frac{S_i}{N_i}\right]_{B_o} \tag{5}$$

If there exists a factor K_ρ relating the peak to rms signal voltage such that

$$s_{o\ r.m.s} = K_\rho\, s_{o\ peak} \tag{6}$$

then the ratio of mean squared signal S_o to mean squared noise N_o may be written

$$\frac{S_o}{N_o} = K_\rho^2 \sin^2\left(\Delta\varphi_p\right) \left[\frac{S_i}{N_i}\right]_{B_o} \tag{7}$$

D.2 Prelimited Product Demodulators

Results similar to those in the preceding section are obtained for a product demodulator preceded by a hard band-pass limiter. The configuration is shown in figure D.2-1. This is the detector of section B.4.2, page B-7, followed by an ideal output filter

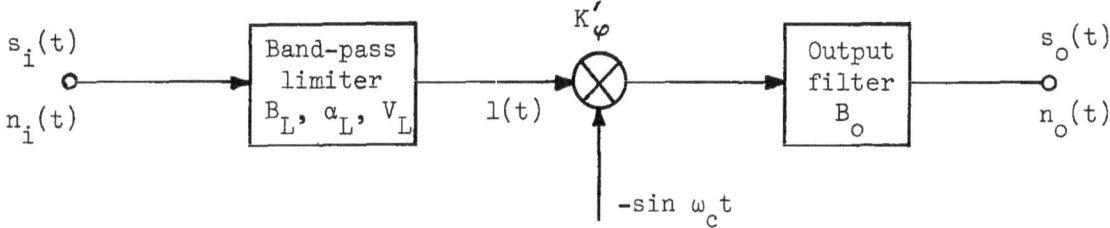

Figure D.2-1.- Demodulator configuration

The ideal output filter has the same characteristics as in D.1, page D-1.

The input signal and noise are taken in the usual form as

$$s_i(t) = A \cos\left[\omega_c t + \varphi_s(t)\right] \qquad (1)$$

$$n_i(t) = x(t) \cos \omega_c t - y(t) \sin \omega_c t \qquad (2)$$

where the signal and noise characteristics are the same as in D.1, page D-1.

For a high input signal-to-noise ratio into the limiter of, say, 10 db, the multiplier signal may be taken from equation B.4.2 (7), page B-9, as approximately

$$m(t) \quad \frac{2K'_\varphi V_L}{\pi} \left[\sin \varphi_s(t) + \frac{y(t)}{A}\right] \qquad (3)$$

For high input signal-to-noise ratios, equation (3) gives output signal-to-noise ratios identical to those for no prelimiting. However, for decreasing input signal-to-noise ratios, limiter effects become pronounced.

Lacking a useful rigorous treatment, the following rough approximation will be made, which highly simplifies the analysis. The multiplier output $m(t)$ will be approximated for all input signal-to-noise ratios by

$$m(t) \cong \frac{2K'_\varphi V_L}{\pi}\left[\alpha_s \sin \varphi_s(t) + \alpha_n \frac{y(t)}{A}\right] \qquad (4)$$

where

α_s = limiter signal suppression factor.

Martin's (ref. 7) approximation for α_s will be used.

$$\alpha_s^2 = \frac{1}{1 + \frac{4}{\pi}\left[\frac{N_i}{S_i}\right]_{B_L}} \qquad (5)$$

where

$\left.\frac{N_i}{S_i}\right|_{B_L}$ = limiter input signal-to-noise ratio in the limiter bandwidth B_L.

It remains to determine the nature of α_n. This may be determined by noting that the limiter is a constant power output device. That is, regardless of the limiter output signal-to-noise ratio, the total output signal plus noise power is constant. This means that regardless of the limit output spectral composition, the total power across the spectrum is constant. Next, it is noted that a product detector is simply a spectral translator. It does not change the nature of the limiter output spectrum, but merely translates it in frequency. Therefore, the total power out of the multiplier is constant. This constancy of multiplier output power P_m will be used to solve for α_n.

$$P_m = 2\left[K'_\varphi \frac{V_L}{\pi}\right]^2 \cong 4\left[\frac{K'_\varphi V_L}{\pi}\right]\left[\frac{\alpha_s^2}{2} + \alpha_n^2 \frac{\overline{y^2(t)}}{A^2}\right] \qquad (6)$$

$$\alpha_n^2 \cong \left[1 - \alpha_s^2\right]\frac{\frac{A^2}{2}}{\overline{y^2(t)}} = \left[1 - \alpha_s^2\right]\left[\frac{S_i}{N_i}\right]_{B_L} \qquad (7)$$

$$\alpha_n^2 \cong \frac{4}{\pi} \alpha_s^2 \tag{8}$$

Using equation (8), it is seen that the signal-to-noise ratio at the multiplier output is given as

$$\frac{S_m}{N_m} = \frac{\frac{\alpha_s^2}{2}}{\alpha_n^2 \frac{\overline{y^2(t)}}{A^2}} = \frac{\pi}{4} \left[\frac{S_i}{N_i}\right]_{B_L} \tag{9}$$

It is apparent that the rough approximation of equations (4) and (8) has given a pessimistic result for output signal-to-noise ratio which is analogous to Davenport's (ref. 4) limiter result at low input signal-to-noise ratios.

Equation (4) is rewritten as

$$m(t) \cong \frac{2K'_\varphi V_L}{\pi} \alpha_s \left[\sin \varphi_s(t) + \sqrt{\frac{4}{\pi}} \frac{y(t)}{A}\right] \tag{10}$$

The signal component and noise spectral density are given separately as

$$s_m(t) = \frac{2K'_\varphi V_L}{\pi} \alpha_s \sin \varphi_s(t) \tag{11}$$

$$\Phi_m(\omega) = \left[\frac{2K'_\varphi V_L}{\pi}\right]^2 \frac{4}{\pi} \alpha_s^2 \frac{\Phi_y(\omega)}{A^2} \tag{12}$$

It should be emphasized that the material presented above in section D.2 is the result of physical reasoning and approximation and is not mathematically rigorous. This material should be applied with care.

APPENDIX E

DEMODULATION WITH MODULATION TRACKING LOOPS

Figure E-1 shows the configuration of a modulation tracking phase-locked loop used to detect frequency modulation.

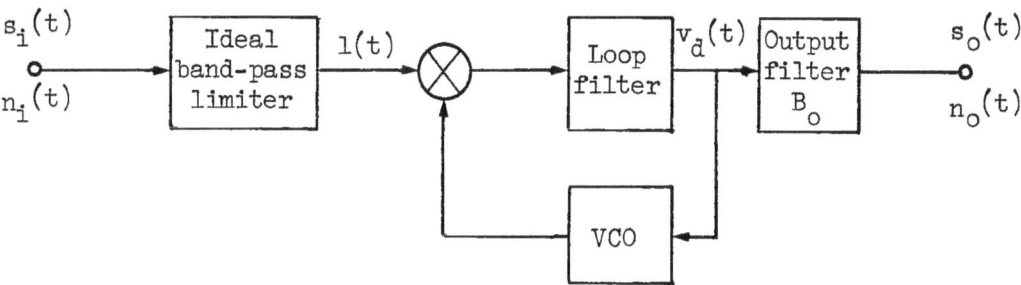

Figure E-1.- Demodulator configuration

The input signal is taken in usual form as

$$s_i(t) = A \cos \left[\omega_c t + \varphi_s(t)\right] \qquad (1)$$

where

$\varphi_s(t)$ = "equivalent" phase modulation of the signal

The input noise is taken as

$$n_i(t) = x(t) \cos \omega_c t - y(t) \sin \omega_c t \qquad (2)$$

The limiter output signal is taken from equation B.3 (4), page B-6, as

$$l(t) = \frac{4}{\pi} V_L \cos \left[\omega_c t + \varphi_i(t)\right] \qquad (3)$$

with $\varphi_i(t)$ identical to $\psi(t)$ of the referenced equation. For sufficiently high limiter signal-to-noise ratio, $\varphi_i(t)$ may be separated into signal phase modulation, $\varphi_s(t)$, and noise phase modulation, $\varphi_\varphi(t)$.

Since a modulation tracking loop is useful only when operating relatively linearly, the assumptions are made that the steady-state modulation tracking error is less than, say, 30° and that the set of linear transfer functions derived in appendix C adequately describe loop operation.

The transfer function of interest is that relating the VCO driving signal $v_d(t)$ to input signal $\varphi_i(t)$. In transform notation, from equation C.2.1 (9), page C-7,

$$\frac{V_D(s)}{\Phi_i(s)} = K_\varphi K_f \left[\frac{s(s-z)}{s^2 + s(K-p) - Kz} \right] \qquad (4)$$

where

$\Phi_i(s)$ = transform of the input signal "equivalent" phase modulation

For signals which are frequency modulated, the transform relation between VCO driving signal and input frequency modulation is given as

$$\frac{V_D(s)}{s\Phi_i(s)} = K_\varphi K_f \left[\frac{s-z}{s^2 + s(K-p) - Kz} \right] \qquad (5)$$

where

$s\Phi_i(s)$ = time integral of the "equivalent" phase modulation or frequency modulation

The asymptotic Bode plot of the steady-state transfer function derived from equation (3) was given in figure C.2.1-1, page C-8, and is reproduced here as figure E-2.

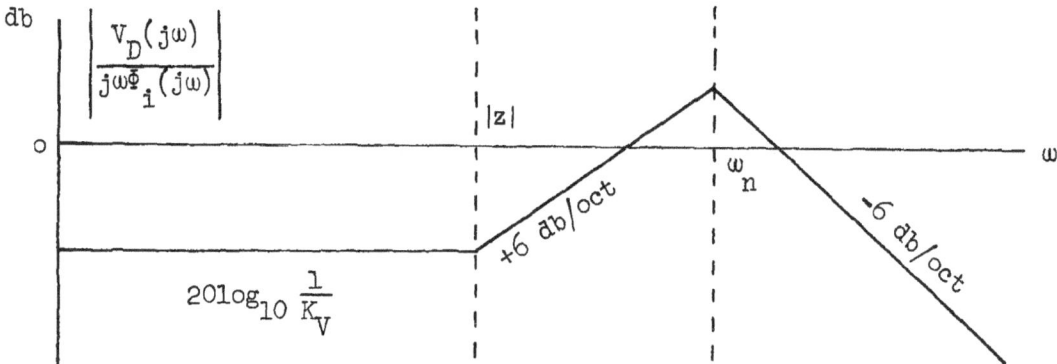

Figure E-2.- Asymptotic Bode plot

The figure shows that the steady-state transfer function is asymptotically flat for frequency modulation from zero frequency out to the region of $\omega = |z|$, the loop filter zero frequency. For analytic simplicity, it may be assumed that most of the signal modulation energy will be of frequency less than $|z|$. This assumption is not strictly necessary, since equalization in the output filter may be employed if the modulation frequencies do extend beyond $|z|$. For modulation satisfying this frequency restriction, the VCO driving signal is given as

$$v_d(t) = \frac{1}{K_V} \dot{\varphi}_i(t) \tag{6}$$

For the assumption of relatively high input signal-to-noise ratio into the limiter and linear loop operation, the VCO driving signal is separable into individual signal and noise components. Then

$$v_d(t) = \frac{1}{K_V} \left[\dot{\varphi}_s(t) + \dot{\varphi}_\varphi(t) \right] \tag{7}$$

The demodulator may be treated for signal and for noise, separately.

E.1 Detection of Sinusoidal Subcarriers and Arbitrary Baseband Modulation

This section treats the demodulation of a carrier frequency modulated by a composite function consisting of a summation of K sinusoidal subcarriers plus some arbitrary baseband function. Figure E-1 applies.

The input signal is taken as

$$s_i(t) = A \cos \left[\omega_c t + \varphi_s(t) \right] \tag{1}$$

where

$\varphi_s(t)$ = carrier phase modulation due to signal alone

The function describing the instantaneous carrier frequency deviation is defined as

$$\dot{\varphi}_s(t) \triangleq f_b(t) + \sum_{i=1}^{K} \Delta\omega_i \cos \left[\omega_i t + \theta_i \right] \tag{2}$$

where

$f_b(t)$ = arbitrary baseband function and

$\Delta\omega_i$ = peak radian frequency deviation of the carrier by the ith subcarrier

The signal portion of the VCO driving function is given by

$$v_{d_s}(t) = \frac{1}{K_V}\left\{f_b(t) + \sum_{i=1}^{K} \Delta\omega_i \cos\left[\omega_i t + \theta_i\right]\right\} \quad (3)$$

The output signal from the output filter (band-pass) for the jth subcarrier is given as

$$s_{o_j}(t) = \frac{1}{K_V} \Delta\omega_j \cos\left[\omega_j t + \theta_j\right] \quad (4)$$

Equation (4) holds whether or not the subcarriers themselves are angle modulated.

The peak output signal from the output filter (low-pass) for baseband modulation is given as

$$s_{ob\;peak} = \frac{1}{K_V} \Delta\omega b_{peak} \quad (5)$$

where

$\Delta\omega b_{peak}$ = peak radian frequency deviation due to the baseband modulation

E.2 Noise Characteristics

Observation of figure C.2.1-1, page C-8, shows that the modulation tracking loop has no finite output noise bandwidth for flat input phase noise. Output filters of the low-pass or band-pass type are used to restrict the output noise. The filters are assumed to have ideal properties, that is, square frequency cut-off characteristics and flat transmission characteristics in the pass-band.

For the assumption that the modulation frequencies are less than $|z|$, then the output noise spectrum is parabolic, or proportional to ω^2 as shown by figure C.2.1-1, page C-8. Therefore, the bandwidths B_o of the output filters are not the equivalent noise bandwidths at the output. These will now be computed.

The noise spectrum at the loop filter output is taken as

$$\Phi_{d_n}(j\omega) = \frac{-(j\omega)^2}{K_V^2} \Phi_\varphi(j\omega) \; ; \; |\omega| < |z| \qquad (1)$$

where

$\Phi_\varphi(j\omega)$ = flat input phase noise spectrum.

Equation E.2 (1) follows from equation E (6), page E-3.

E.2.1 Low-pass Output Filter

The low-pass output filter is taken to have an amplitude transmission coefficient of unity and physical bandwidth B_o cps which corresponds to radian bandwidth

$$\Delta\omega_o = 2\pi B_o \qquad (1)$$

The equivalent noise bandwidth B_e is defined as that bandwidth having a transmission constant of unity which passes the same noise power from flat input spectral density $\Phi_\varphi(j\omega)$ as is actually present in the output. Equating noise powers, we have

$$\frac{1}{2\pi j} \frac{1}{K_V^2} |\Phi_\varphi| j2\Delta\omega_e = \frac{1}{2\pi j} \int_{-j\Delta\omega_o}^{j\Delta\omega_o} -\frac{1}{K_V^2}(j\omega)^2 \Phi_\varphi(j\omega) d(j\omega) \qquad (2)$$

where

$$\Delta\omega_e = 2\pi B_e \qquad (3)$$

then

$$j2\Delta\omega_e = -\int_{-j\Delta\omega_o}^{j\Delta\omega_o} (j\omega)^2 d(j\omega) \qquad (4)$$

$$\Delta\omega_e = \frac{(\Delta\omega_o)^3}{3} \qquad (5)$$

$$B_e = \frac{(2\pi)^2}{3} B_o^3 \tag{6}$$

The output noise power is seen to be

$$N_o = \frac{1}{2\pi}\left(\frac{1}{K_V}\right)^2 |\Phi_\varphi| 2\Delta\omega_e \tag{7}$$

$$N_o = \frac{1}{2\pi}\left(\frac{1}{K_V}\right)^2 |\Phi_\varphi| \frac{2}{3}(\Delta\omega_o)^3 \tag{8}$$

$$N_o = \frac{2}{3}\left(\frac{1}{K_V}\right)^2 (2\pi)^2 B_o^3 |\Phi_\varphi| \tag{9}$$

E.2.2 Band-pass Output Filter

The band-pass filter is taken to have an amplitude transmission coefficient of unity and physical one-sided bandwidth of B_o cps which is symmetric to a frequency of f_m cps.

In radian notation,

$$\Delta\omega_o = 2\pi B_o$$
$$\omega_m = 2\pi f_m \tag{1}$$

Equating noise powers give

$$\frac{2}{2\pi j}\int_{j\left(\omega_m - \frac{\Delta\omega_o}{2}\right)}^{j\left(\omega_m + \frac{\Delta\omega_o}{2}\right)} -(j\omega)^2\left(\frac{1}{K_V}\right)^2 \Phi_\varphi(j\omega) d(j\omega) = \frac{1}{2\pi j}\left(\frac{1}{K_V}\right)^2 |\Phi_\varphi| j2\Delta\omega_e \tag{2}$$

Where

$\Delta\omega_e$ = equivalent noise bandwidth at the output

$$-2 \int_{j\left(\omega_m - \frac{\Delta\omega_o}{2}\right)}^{j\left(\omega_m + \frac{\Delta\omega_o}{2}\right)} (j\omega)^2 d(j\omega) = j2\Delta\omega_e \qquad (3)$$

$$\Delta\omega_e = \frac{1}{3}\left[\left(\omega_m + \frac{\Delta\omega_o}{2}\right)^3 - \left(\omega_m - \frac{\Delta\omega_o}{2}\right)^3\right] \qquad (4)$$

$$\Delta\omega_e = \Delta\omega_o\left[\omega_m^2 + \frac{1}{12}\left(\Delta\omega_o\right)^2\right] \qquad (5)$$

$$B_e = (2\pi)^2\left[f_m^2 + \frac{B_o^2}{12}\right] B_o \qquad (6)$$

The output noise power due to a two-sided equivalent phase noise input spectral density $\Phi_\varphi(j\omega)$ is given as

$$N_o = \frac{1}{2\pi}\left(\frac{1}{K_V}\right)^2 \left|\Phi_\varphi\right| 2\Delta\omega_e \qquad (7)$$

$$N_o = \frac{1}{2\pi}\left(\frac{1}{K_V}\right)^2 \left|\Phi_\varphi\right| 2\Delta\omega_o \left[\omega_m^2 + \frac{1}{12}\left(\Delta\omega_o\right)^2\right] \qquad (8)$$

or

$$N_o = \left(\frac{1}{K_V}\right)^2 \left|\Phi_\varphi\right| 2(2\pi)^2 \left[f_m^2 + \frac{B_o^2}{12}\right] B_o \qquad (9)$$

It is seen that from equation (5) an approximation may be made.

$$\Delta\omega_e \cong \omega_m^2 \Delta\omega_o \; ; \; \omega_m^2 \gg \frac{1}{12}(\Delta\omega_o)^2 \qquad (10)$$

or

$$B_e = (2\pi)^2 f_m^2 B_o \; ; \; f_m^2 \gg \frac{1}{12} B_o^2 \qquad (11)$$

This approximation is accurate to within about 5 percent for $\Delta\omega_o$ approaching $\frac{3}{4}\omega_m$. This approximation is essentially the same as assuming flat noise with spectral density

$$\Phi_{n_o}(j\omega) = \omega_m^2 \Phi_\varphi(j\omega) \qquad (12)$$

across the bandwidth B_o when B_o is much less than the center frequency f_m. Approximately

$$N_o \cong \frac{1}{2\pi}\left(\frac{1}{K_V}\right)^2 |\Phi_\varphi| 2\Delta\omega_o \omega_m^2 \qquad (13)$$

or

$$N_o \cong \left(\frac{1}{K_V}\right)^2 |\Phi_\varphi| 2(2\pi)^2 B_o f_m^2 \qquad (14)$$

E.3 Output Signal-to-noise Ratios

The results of the prior sections are now used to obtain output signal-to-noise ratios for two cases: arbitrary baseband modulation with a low-pass filter, and an individual subcarrier with a band-pass filter.

E.3.1 Subcarrier and Band-pass Filter

We may now determine the signal-to-noise ratio out of the assumed square band-pass filter of bandwidth B_o for the jth subcarrier, where the input signal is assumed to be frequency modulated by the sum of K subcarriers plus arbitrary baseband modulation. The input noise is white, band-limited, and Gaussian.

From equation E.1 (4), page E-4, the output signal for the jth subcarrier is

$$s_{oj}(t) = \frac{1}{K_V} \Delta\omega_j \cos\left[\omega_j t + \varphi_j\right] \qquad (1)$$

The output signal power is then

$$S_{oj} = \frac{1}{K_V^2} \frac{\Delta\omega_j^2}{2} \qquad (2)$$

From equation E.2.2 (9), page E-7, the output noise power is

$$N_o = \frac{1}{K_V^2} \left|\Phi_\varphi\right| 2(2\pi)^2 \left[f_j^2 + \frac{B_o^2}{12}\right] B_o$$

where

f_j = center frequency of the band-pass filter

The output signal-to-noise ratio is then

$$\frac{S_{oj}}{N_o} = \frac{\dfrac{1}{K_V^2} \dfrac{\Delta\omega_j^2}{2}}{\dfrac{1}{K_V^2} \left|\Phi_\varphi\right| 2(2\pi)^2 \left[f_j^2 + \dfrac{B_o^2}{12}\right] B_o} \qquad (4)$$

$$\frac{S_{oj}}{N_o} = \frac{\dfrac{\Delta\omega_j^2}{2}}{(2\pi)^2 \left[f_j^2 + \dfrac{B_o^2}{12}\right]} \left[\frac{\dfrac{A^2}{2}}{\left|\Phi_{ni}\right| 2B_o}\right] \qquad (5)$$

where

$$\Phi_\varphi(\omega) \equiv \frac{\Phi_{ni}(\omega)}{\dfrac{A^2}{2}} \qquad (6)$$

The second bracketed quantity in equation (5) is seen to be the input signal-to-noise ratio computed in a physical bandwidth B_o. Then

$$\frac{S_{oj}}{N_o} = \frac{1}{2} \frac{\Delta f_j^2}{\left[f_j^2 + \frac{B_o^2}{12}\right]} \left[\frac{S_i}{N_i}\right]_{B_o} \tag{7}$$

where

Δf_j = peak cyclic frequency deviation of the carrier by the jth subcarrier

Approximately

$$\frac{S_{oj}}{N_o} \cong \frac{1}{2} \left[\frac{\Delta f_j}{f_j}\right]^2 \left[\frac{S_i}{N_i}\right]_{B_o} \tag{8}$$

for

$$1 \gg \frac{1}{12} \left[\frac{B_o}{f_j}\right]^2 \tag{9}$$

E.3.2 Baseband Modulation and Low-pass Filter

Proceeding as in E.3.1, the peak output signal from equation E.1 (5), page E-4, is

$$s_{ob\ peak} = \frac{1}{K_V} \Delta\omega_{b\ peak} \tag{1}$$

The peak-squared signal is

$$\left[s_{ob\ peak}\right]^2 = \frac{1}{K_V^2} \Delta\omega_{b\ peak}^2 \tag{2}$$

From equation E.2.1 (9), page E-6, the output noise power is

$$N_o = \frac{2}{3} \left[\frac{1}{K_V}\right]^2 (2\pi)^2 B_o^3 \left|\Phi_\varphi\right| \tag{3}$$

The ratio of peak-squared signal to mean-squared noise is then

$$\frac{\left[S_{ob\ peak}\right]^2}{N_o} = \frac{\Delta\omega b^2_{peak}}{\frac{2}{3}(2\pi)^2 B_o^3 |\Phi_\varphi|} \tag{4}$$

$$\frac{\left[S_{ob\ peak}\right]^2}{N_o} = 3\left[\frac{\Delta fb_{peak}}{B_o}\right]^2 \left[\frac{\frac{A^2}{2}}{|\Phi_{ni}|\ 2B_o}\right] \tag{5}$$

where

Δfb_{peak} = peak cyclic frequency deviation of the carrier by the baseband modulation.

The second bracketed quantity is seen to be the input signal-to-noise ratio computed in a physical bandwidth B_o. Then

$$\frac{\left[S_{ob\ peak}\right]^2}{N_o} = 3\left[\frac{\Delta fb_{peak}}{B_o}\right]^2 \left[\frac{S_i}{N_i}\right]_{B_o} \tag{6}$$

If a specification factor K_p exists, relating the peak to rms value of the baseband modulation, such that

$$fb_{r.m.s.} = K_p fb_{peak} \tag{7}$$

then the mean-squared output signal-to-noise ratio may be formed as

$$\frac{S_{ob}}{N_o} = 3K_p^2 \left[\frac{\Delta fb_{peak}}{B_o}\right]^2 \left[\frac{S_i}{N_i}\right]_{B_o} \tag{8}$$

E-11

APPENDIX F

SPECIALIZED DETECTORS

F.1 Range Clock Receiver and Code Correlator

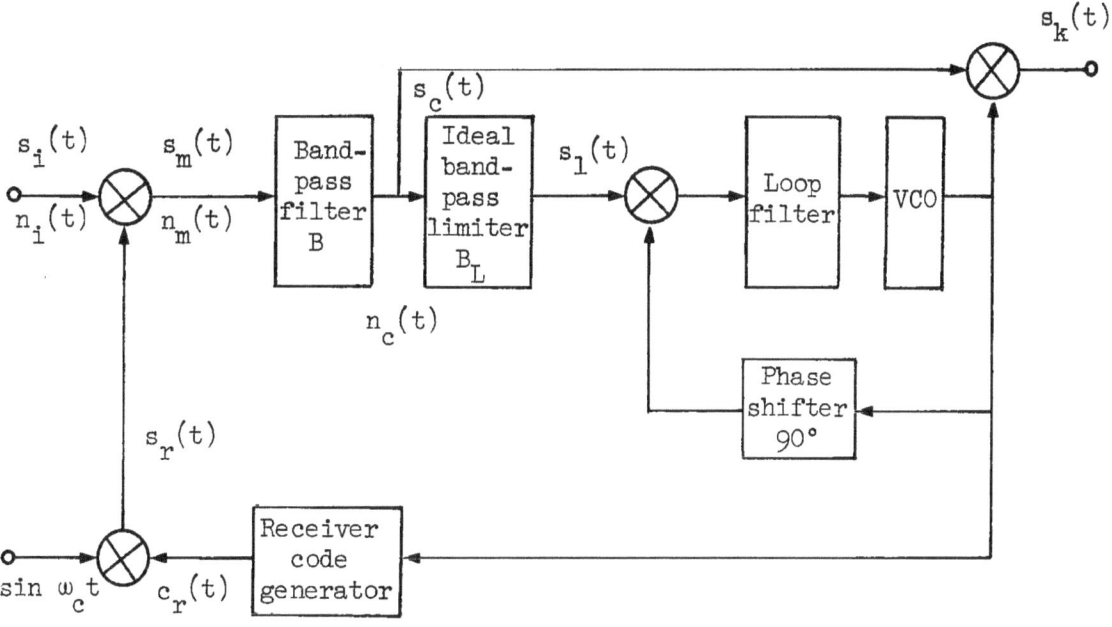

Figure F.1-1.- Range clock receiver

The above figure shows the block diagram of the circuitry which recovers the received range clock signal and generates the range code correlation signal. The physical operation of this circuitry has been treated in volume I of this series.

$s_i(t) + n_i(t)$ = the IF input signal plus noise

$c_r(t)$ = receiver code having only values of ±1

$s_r(t)$ = product of $c_r(t)$ and the IF reference signal

$s_m(t) + n_m(t)$ = signal plus noise out of the multiplier

$s_c(t)$ = correlator driving signal

$s_k(t)$ = correlator output signal

$s_l(t)$ = clock loop driving signal

The operation of this device will be treated for an input signal, phase modulated by a transmitted ranging code $c_t(t)$, plus K subcarriers. The input noise is assumed Gaussian, flat, and band-limited to B_i cps.

F.1.1 Signal Treatment

$$s_i(t) = A \cos \left\{ \omega_c t + \Delta\varphi_r c_t(t) + \sum_{i=1}^{K} \Delta\varphi_i \sin \left[\omega_i t + \varphi_i(t) \right] \right\} \quad (1)$$

where $s_i(t)$ has the same properties treated in appendix A.2.

$$s_r(t) = -c_r(t) \sin \omega_c t \quad (2)$$

$$s_m(t) = s_i(t) \, s_r(t) \quad (3)$$

$$s_m(t) = -A \, c_r(t) \sin \omega_c t \, \cos \left\{ \omega_c t + \Delta\varphi_r c_t(t) + \sum_{i=1}^{K} \Delta\varphi_i \sin \left[\omega_i t + \varphi_i(t) \right] \right\}$$

$$(4)$$

It is assumed that the multiplier produces only the difference frequency term.

then

$$s_m(t) = \frac{A}{2} c_r(t) \sin \left\{ \Delta\varphi_r c_t(t) + \sum_{i=1}^{K} \Delta\varphi_i \sin \left[\omega_i t + \varphi_i(t) \right] \right\} \quad (5)$$

Equation (5) may be expanded as in section D.1.1.

$$s_m(t) = \frac{A}{2} \sin\left(\Delta\varphi_r\right) c_r(t) c_t(t) \sum_{n_1=-\infty}^{\infty} \cdots \sum_{n_k=-\infty}^{\infty} \prod_{i=1}^{K} \left[J_{n_i}\left(\Delta\varphi_i\right)\right] \cdot$$

$$\cdot \cos\left\{\sum_{i=1}^{K} n_i\left[\omega_i t + \varphi_i(t)\right]\right\} + \frac{A}{2} \cos\left(\Delta\varphi_r\right) c_r(t) \sum_{n_1=\infty}^{\infty} \cdots \sum_{n_k=\infty}^{\infty}$$

$$\prod_{i=1}^{K} \left[J_{n_i}\left(\Delta\varphi_i\right)\right] \sin\left\{\sum_{i=1}^{K} n_i\left[\omega_i t + \varphi_i(t)\right]\right\}$$

(6)

The first term in equation (6) is the desired term. The second term is the interference. Equation (6) holds whether or not the subcarriers themselves are angle modulated. For the digital logic employed by this system the product of the two analog code waveforms $c_r(t)$ and $c_t(t)$ corresponds to the Boolean modulo two addition, or "exclusive or", of the two codes C_r and C_t^*. The receiver code output is programable in seven steps. At each step, the correlation between the two codes changes. This is shown in table I. It can be shown, either algebraically or with a truth table, that the "exclusive or" of the two codes in program state P7 is identically clock, Cl. This means that

$$c_{r_7}(t) c_t(t) = cl(t) \tag{7}$$

where

cl(t) = square wave of unit amplitude, having the clock frequency ω_{cl}.

---

*For a detailed physical explanation of the ranging equipment, see volume I of this series.

TABLE F.1.1-I.- PROGRAM STATE VERSUS CORRELATION

$C_T = C1 \oplus X(abvbcvac)$

Transmitter code					Bit length
Program state	Receiver code	Component acquired	Initial correlation	Final correlation	
P1	0	C1	0	50%	0
P2					
P3	$\overline{X}a$	X	25%	50%	341
P4	$\overline{X}a$	a	50%	75%	341
P5	$\overline{X}b$	b	50%	75%	693
P6	$\overline{X}c$	c	50%	75%	1 397
P7	$\overline{X}(abvbcvac)$	check a, b, c	75%	100%	2 728 341

Code component	Bit length
X	11
a	31
b	63
c	127
C1	2

In terms of Fourier series,

$$c_{r_7}(t)c_t(t) = \frac{2}{\pi} \sum_{p=1}^{\infty} \frac{1}{p}\left[1 - \cos(p\pi)\right] \sin p\omega_{cl} t \quad (8)$$

where

$$\omega_{cl} = \frac{\pi}{R}$$

R = clock bit period (9)

Then the correlation driving signal for state P7 is

$$s_{c_7}(t) = \frac{2A}{\pi} \sin\left(\Delta\varphi_r\right) \prod_{i=1}^{K} J_o\left(\Delta\varphi_i\right) \sin \omega_{cl} t \quad (10)$$

where it has been assumed that the band-pass filter having bandwidth B passes only the fundamental sinusoidal component of the clock square wave.

F.1.2 Noise Treatment

The input noise is taken in the usual form as

$$n_i(t) = x(t) \cos \omega_c t - y(t) \sin \omega_c t \quad (1)$$

having an input noise spectral density $\Phi_{ni}(\omega)$, band-limited to B_i. The noise term from the multiplier is

$$n_m(t) = n_i(t)\left[-c_r(t) \sin \omega_c t\right] \quad (2)$$

$$n_m(t) \cong \frac{1}{2} c_r(t) y(t) \quad (3)$$

We are now interested in obtaining $\Phi_m(\omega)$, the spectral density of the noise term out of the multiplier. We make the assumption that $c_r(t)$ and $y(t)$ may be represented as sample functions of independent random processes and that

$$\Phi_m(\omega) = \frac{1}{4} \Phi_{cr}(\omega) * \Phi_y(\omega) \tag{4}$$

where

$$\Phi_y(\omega) = 2|\Phi_{ni}| \quad ; \quad |\omega| < 2\pi \frac{B_i}{2} \triangleq \frac{\Delta\omega_i}{2} \tag{5}$$

$$= 0 \quad ; \text{ all other } \omega$$

From Titsworth and Welch (ref. 10) we approximate the spectral density of $c_r(t)$ in its program states greater than P2, as that of a Markov sequence.

Then

$$\Phi_{cr}(\omega) = \frac{1}{R} \frac{\sin^2\left(\frac{\omega}{2R}\right)}{\left(\frac{\omega}{2R}\right)^2} \tag{6}$$

where

$R = $ bit rate of $c_r(t)$

Now,

$$\Phi_m(\omega) = \frac{1}{4}\left[\frac{1}{2\pi}\int_{-\infty}^{\infty} \Phi_{cr}(y)\Phi_y(\omega - y)dy\right] \tag{7}$$

Since

$$\Phi_y(\omega - y) = 2|\Phi_{Ni}| \quad ; \quad -\frac{\Delta\omega_i}{2} < \omega - y < \frac{\Delta\omega_i}{2}$$

$$= 0 \quad ; \quad |\omega - y| > \frac{\Delta\omega_i}{2} \tag{8}$$

then

$$\Phi_y(\omega - y) = 2|\Phi_{Ni}| \quad ; \quad \omega - \frac{\Delta\omega_i}{2} < y < \omega + \frac{\Delta\omega_i}{2}$$

$$= 0 \quad ; \text{ all other } y \tag{9}$$

It follows that

$$\Phi_m(\omega) = \frac{1}{8\pi} \int_{\omega - \frac{\Delta\omega_i}{2}}^{\omega + \frac{\Delta\omega_i}{2}} 2\Phi_{cr}(y) \left|\Phi_{ni}\right| dy \qquad (10)$$

$$\Phi_m(\omega) = \frac{\left|\Phi_{ni}\right|}{4\pi} \int_{\omega - \frac{\Delta\omega_i}{2}}^{\omega + \frac{\Delta\omega_i}{2}} \Phi_{cr}(y) dy \qquad (11)$$

$$\Phi_m(\omega) = \left|\frac{\Phi_{ni}}{4\pi}\right| \int_{\omega - \frac{\Delta\omega_i}{2}}^{\omega + \frac{\Delta\omega_i}{2}} \frac{1}{R} \frac{\sin^2\left(\frac{y}{2R}\right)}{\left(\frac{y}{2R}\right)^2} dy \qquad (12)$$

Let

$$\frac{y}{2R} = x \; ; \; y = 2Rx \; ; \; dy = 2Rdx \qquad (13)$$

then

$$\Phi_m(\omega) = \left|\frac{\Phi_{ni}}{4\pi}\right| \int_{\frac{\omega - \frac{\Delta\omega_i}{2}}{2R}}^{\frac{\omega + \frac{\Delta\omega_i}{2}}{2R}} \frac{1}{R} \frac{\sin^2 x}{x^2} 2Rdx \qquad (14)$$

$$\Phi_m(\omega) = \left|\frac{\Phi_{ni}}{2\pi}\right| \int_{\frac{\omega - \frac{\Delta\omega_2}{2}}{2R}}^{\frac{\omega + \frac{\Delta\omega_i}{2}}{2R}} \frac{\sin^2 x}{x^2} dx \qquad (15)$$

F-7

Now

$$\int_{x_1}^{x_2} \frac{\sin^2 x}{x^2} dx = \frac{1}{2} \left[\int_{x_1}^{x_2} \frac{dx}{x^2} - \int_{x_1}^{x_2} \frac{\cos 2x}{x^2} dx \right] \quad (16)$$

$$\int_{x_1}^{x_2} \frac{\sin^2 x}{x^2} dx = \frac{1}{2} \left[\int_{x_1}^{x_2} \frac{dx}{x^2} + \left. \frac{\cos 2x}{x} \right|_{x_1}^{x_2} + 2 \int_{x_1}^{x_2} \frac{\sin 2x}{x} dx \right] \quad (17)$$

$$\int_{x_1}^{x_2} \frac{\sin^2 x}{x^2} dx = \frac{1}{2} \left\{ \frac{1}{x_1} - \frac{1}{x_2} + \frac{\cos 2x_2}{x_2} - \frac{\cos 2x_1}{x_1} + 2 \left[S_i(2x_2) - S_i(2x_1) \right] \right\} \quad (18)$$

where

$$S_i(z) \equiv \int_0^z \frac{\sin x}{x} dx \quad (19)$$

$$\Phi_m(\omega) = \left|\frac{\Phi_{ni}}{4\pi}\right| \left\{ \frac{1}{\frac{\omega - \frac{\Delta\omega_i}{2}}{2R}} - \frac{1}{\frac{\omega + \frac{\Delta\omega_i}{2}}{2R}} + \frac{\cos\left(\frac{\omega + \frac{\Delta\omega_i}{2}}{R}\right)}{\frac{\omega + \frac{\Delta\omega_i}{2}}{2R}} \right.$$

$$\left. - \frac{\cos\left(\frac{\omega - \frac{\Delta\omega_i}{2}}{R}\right)}{\frac{\omega - \frac{\Delta\omega_i}{2}}{2R}} + 2\left[S_i\left(\frac{\omega + \frac{\Delta\omega_i}{2}}{R}\right) - S_i\left(\frac{\omega - \frac{\Delta\omega_i}{2}}{R}\right)\right] \right\} \quad (20)$$

and

$$\Phi_m(\omega) = \left|\Phi_{ni}\right| \frac{R}{2\pi} \left\{ \frac{\Delta\omega_i}{\omega^2 - \frac{\Delta\omega_i^2}{4}} + \frac{\cos\left(\frac{\omega + \frac{\Delta\omega_i}{2}}{R}\right)}{\omega + \frac{\Delta\omega_i}{2}} - \frac{\cos\left(\frac{\omega - \frac{\Delta\omega_i}{2}}{R}\right)}{\omega - \frac{\Delta\omega_i}{2}} \right.$$

$$\left. + \frac{1}{R}\left[S_i\left(\frac{\omega + \frac{\Delta\omega_i}{2}}{R}\right) - S_i\left(\frac{\omega - \frac{\Delta\omega_i}{2}}{R}\right)\right] \right\} \quad (21)$$

Equation (21) is the general expression for the noise spectral density at the output of the multiplier. We are interested in evaluating this spectral density at the center frequency of the narrow band-pass filter. We will then assume $\Phi_m(\omega)$ to be flat across this narrow bandwidth.

The frequency of interest is

$$\omega = \pi R \quad (22)$$

The assumption is also made that the input bandwidth may be limited to

$$\Delta\omega_i = 10\pi R \tag{23}$$

Using the assumptions stated above, the noise spectral density at the output of the band-pass filter B may be written as

$$\Phi_c(\omega) = \Phi_m(\omega)\Big|_{\omega=\pi R} \tag{24}$$

$$\Phi_c(\omega) = |\Phi_{ni}| \frac{R}{2\pi} \left\{ -\frac{10\pi R}{24\pi^2 R^2} + \frac{\cos(6\pi)}{6\pi R} - \frac{\cos(-4\pi)}{-4\pi R} \right.$$
$$\left. + \frac{1}{R}\left[S_i(6\pi) - S_i(-4\pi)\right] \right\} \tag{25}$$

$$\Phi_c(\omega) = \frac{|\Phi_{ni}|}{2\pi} \cdot \left\{ -\frac{10}{24\pi} + \frac{1}{6\pi} + \frac{1}{4\pi} + S_i(18.85) + S_i(12.55) \right\} \tag{26}$$

$$\Phi_c(\omega) = \frac{3.06}{2\pi} |\Phi_{ni}| \tag{27}$$

$$\Phi_c(\omega) = 0.488 |\Phi_{ni}| \tag{28}$$

Now, the noise power N_c at the output of the band-pass filter in the bandwidth B is

$$N_c = |\Phi_c| 2B \tag{29}$$

$$N_c = .976 |\Phi_{ni}| B \tag{30}$$

F.1.3 Signal-to-noise Ratios

From equations F.1.1 (10) and F.1.2 (27) we may obtain the signal-to-noise ratio at the output of the band-pass filter of bandwidth B for program state P_7 as

$$\left[\frac{S_c}{N_c}\right]_B\Bigg|_{P_7} = -\frac{2\frac{A^2}{\pi^2}\sin^2(\Delta\varphi_r)\prod_{i=1}^{K} J_o^{\,2}(\Delta\varphi_i)}{\frac{3.06}{2\pi}\left|\Phi_{ni}\right|2B} \tag{1}$$

$$\left[\frac{S_c}{N_c}\right]_B\Bigg|_{P_7} = \frac{2.62}{\pi}\sin^2(\Delta\varphi_r)\prod_{i=1}^{K} J_o^{\,2}(\Delta\varphi_i)\left[\frac{\frac{A^2}{2}}{\left|\Phi_{ni}\right|2B}\right] \tag{2}$$

Equation (2) relates to the input signal-to-noise ratio computed in a bandwidth B as

$$\left[\frac{S_c}{N_c}\right]_B\Bigg|_{P_7} = \frac{2.62}{\pi}\sin^2(\Delta\varphi_r)\prod_{i=1}^{K} J_o^{\,2}(\Delta\varphi_i)\left[\frac{S_i}{N_i}\right]_B \tag{3}$$

For program state P_7, the product $c_r(t)\,c_t(t)$ is a square wave clock signal 100 percent of the time. For other program states $c_r(t)\,c_t(t)$ is a square wave which reverses phase some percent of the time, on the average, depending on the percent correlation of C_r with C_t. This behavior, coupled with the filtering action of the band-pass filter, is interpreted as causing the amplitude of $s_c(t)$ to be proportional to the average amount of time $c_r(t)\,c_t(t)$ is constant phase clock, or proportional to the correlation of C_r with C_t. A proportionality factor $\sqrt{L_K}$, normalized to the P_7 value of amplitude of $s_c(t)$, is employed to account for the variation. $\sqrt{L_K}$ has maximum value of 1.0 in state P_7 and minimum value of 0.25 in state P_3. L_K is defined as a correlation loss.

Equation (3) may be generalized as

$$\left[\frac{S_c}{N_c}\right]_B = \frac{2.62}{\pi}L_K\sin^2(\Delta\varphi_r)\prod_{i=1}^{K} J_o^{\,2}(\Delta\varphi_i)\left[\frac{S_i}{N_i}\right]_B \tag{4}$$

From equation A.3 (6), page A-8, it may be determined that the receiver input signal-to-noise ratio for the range code component <u>only</u>, $\left[\dfrac{S_{ir}}{N_{ir}}\right]_B$ is related to the input signal-to-noise ratio for the total carrier by

$$\left[\frac{S_{ir}}{N_{ir}}\right]_B = \sin^2\left(\Delta\varphi_r\right) \prod_{i=1}^{K} J_o^2\left(\Delta\varphi_i\right) \left[\frac{S_i}{N_i}\right]_B \tag{5}$$

It is seen that there has been an effective signal loss due to the effects of the receiver code on the input noise, given by the factor $\dfrac{2.62}{\pi}$ or 0.835.

Equation (4) may be generalized as

$$\left[\frac{S_c}{N_c}\right]_B = L_D L_K \sin^2\left(\Delta\varphi_r\right) \prod_{i=1}^{K} J_o^2\left(\Delta\varphi_i\right) \left[\frac{S_i}{N_i}\right]_B \tag{6}$$

where

L_D = .835 is defined as detection loss.

F.1.4 Receiver Threshold

The threshold of the range clock receiver is the threshold of the clock loop in the receiver. The threshold treatment of the clock loop is that for a modulation restrictive loop as given in section C.4. The signal-to-noise ratio used for threshold computations is that given by equation F.1.3 (6) above, except that it is computed in the clock loop noise bandwidth. Thus,

$$\left[\frac{S_c}{N_c}\right]_{B_N} = L_D L_K \sin^2\left(\Delta\varphi_r\right) \prod_{i=1}^{K} J_o^2\left(\Delta\varphi_i\right) \left[\frac{S_i}{N_i}\right]_{B_N} \tag{1}$$

where $\left[\dfrac{S_i}{N_i}\right]_{B_N}$ is the total carrier-to-noise ratio at the input to the range clock receiver, computed in the bandwidth B_N. The value of L_D is 0.835. A worst case value of L_K is -12 decibels.

F.1.5 Range Code Acquisition Time

It is desirable to relate the time for acquiring the ranging code to the signal and noise parameters at the input to the range clock receiver, shown in figure F.1-1, page F-1. Acquisition time is defined as being the total time to obtain indications of correlation between the locally generated code components and the received code for a given probability or error, given prior acquisition of the range clock signal. Rapid clock acquisition is assured for the clock loop signal-to-noise ratio sufficiently high.

The code acquisition process has been physically described elsewhere (ref. 11). The process is known as "maximum likelihood acquisition," and has been treated by Easterling (ref. 12). With a slight modification Easterling's treatment may be applied directly to the range clock receiver and code correlator. The modification is that here the energy per bit is given by the difference in correlation levels at the output of the correlator and not by the levels themselves.

From equations F.1.1 (10), page F-5, and F.1.3 (6), page F-12, it is seen that the signal into the correlator is

$$s_c(t) = \frac{2A}{\pi} \sqrt{L_K} \sin(\Delta\varphi_r) \prod_{i=1}^{K} J_o(\Delta\varphi_i) \sin \omega_{cl} t \qquad (1)$$

The reference signal from the clock loop VCO has sine phase. The significant correlator output term, the difference term, is then taken as

$$s_K(t) = \frac{A}{\pi} \sqrt{L_K} \sin(\Delta\varphi_r) \prod_{i=1}^{K} J_o(\Delta\varphi_i) \qquad (2)$$

The desired output signal $s_o(t)$ is the change of $s_K(t)$ when a code component is acquired. The change is always positive (ref. 11), therefore, $s_o(t)$ is always positive. $\sqrt{L_K}$ has only values 0.25, 0.50, 0.75, or 1.00 (ref. 11). Acquisition of a component is signaled by $\sqrt{L_K}$ increasing from its initial value, say 0.50, to the next higher value, say 0.75. Therefore, the effective output signal may be written as

$$s_o(t) = \frac{A}{\pi} \left[\sqrt{L_{K_2}} - \sqrt{L_{K_1}} \right] \sin(\Delta\varphi_r) \prod_{i=1}^{K} J_o(\Delta\varphi_i) \qquad (3)$$

where

$\sqrt{LK_2}$ = correlation value after acquisition of a component

$\sqrt{LK_1}$ = value before acquisition of a component

Therefore,

$$s_o(t) = \frac{A}{4\pi} \sin(\Delta\varphi_r) \prod_{i=1}^{K} J_o(\Delta\varphi_i) \qquad (4)$$

The signal power at the output of the correlator is then

$$S_o = \frac{A^2}{16\pi^2} \sin^2(\Delta\varphi_r) \prod_{i=1}^{K} J_o^2(\Delta\varphi_i) \qquad (5)$$

The magnitude of the assumed flat noise spectral density at the output of the correlator is

$$\left|\Phi_o\right| = \frac{1}{2}\left|\Phi_c\right| \qquad (6)$$

and is related to the spectral density at the receiver input through equation F.1.2 (27), page F-10, as

$$\left|\Phi_o\right| = \frac{3.06}{4\pi}\left|\Phi_{ni}\right| \qquad (7)$$

The ratio of output signal power to noise spectral density is next obtained as

$$\frac{S_o}{\left|\Phi_o\right|} = \frac{\dfrac{A^2}{16\pi^2} \sin^2(\Delta\varphi_r) \prod_{i=1}^{K} J_o^2(\Delta\varphi_i)}{\dfrac{3.06}{4\pi}\left|\Phi_{ni}\right|} \qquad (8)$$

$$\frac{S_o}{\left|\Phi_o\right|} = \left[\frac{A^2}{2}\right]\left[\frac{1}{2 \times 3.06\,\pi}\right]\left[\frac{\sin^2(\Delta\varphi_r) \prod_{i=1}^{K} J_o^2(\Delta\varphi_i)}{\left|\Phi_{ni}\right|}\right] \qquad (9)$$

$$\frac{S_o}{|\Phi_o|} = \frac{1}{6.12\,\pi} \sin^2(\Delta\varphi_r) \prod_{i=1}^{K} J_o^2(\Delta\varphi_i) \frac{S_i}{|\Phi_{ni}|} \qquad (10)$$

The total code acquisition time for this system, assuming prior clock acquisition, may be broken into two parts: the integration time required to make a decision within the assigned error probability on a maximum likelihood basis, and the built-in machine delay time. The following definitions are made:

T_a = total code acquisition time

T_m = machine delay time between trial correlations

T_i = integration time per trial correlation

W_i = the ith code component

P_i = period, in elements of the ith component W_i

$N_i = \log_2 P_i$ = number of information bits in W_i

1 = subscript designating longest component W_1

T_1 = integration time per information bit in W_1

T_m, the machine delay time, is a built-in fixed parameter of the ranging digital circuitry. T_i is an implicit function of the ratio of correlation signal power to noise spectral density, probability of error, and information content N_i of the code component W_i. The digital ranging system is implemented such that T_i is fixed during any single range code acquisition. Therefore, T_i must be fixed to accommodate the code component W_1 of greatest length, which has the highest information content N_1.

then

$$T_i = N_1 T_1 \qquad (11)$$

The integration is performed sequentially on each element of a code component, and sequentially on the components themselves. Therefore, the total acquisition time may be written as

$$T_a = T_m \left[\Sigma P_i - 1 \right] + T_1 N_1 \Sigma P_i \qquad (12)$$

Since in usable cases the sum of the component periods is much greater than unity, the approximation holds that

$$T_a \cong \Sigma P_i \left[T_1 N_1 + T_m \right] \qquad (13)$$

The integration time per information bit may be solved for, as

$$T_i = \frac{1}{N_1} \left[\frac{T_a}{\Sigma P_i} - T_m \right] \qquad (14)$$

Easterling's (ref. 12) figure is reproduced here as figure F.1.5-1 with the abscissa units relabeled to conform to this notation. For a given error probability and information content of the longest code component, a value of $T_1 \frac{S_o}{|\Phi_o|}$ may be read directly from the curve. Given a requirement for T_1, stemming from a requirement for T_a, the required output value $\frac{S_o}{|\Phi_o|}$ may be inferred.

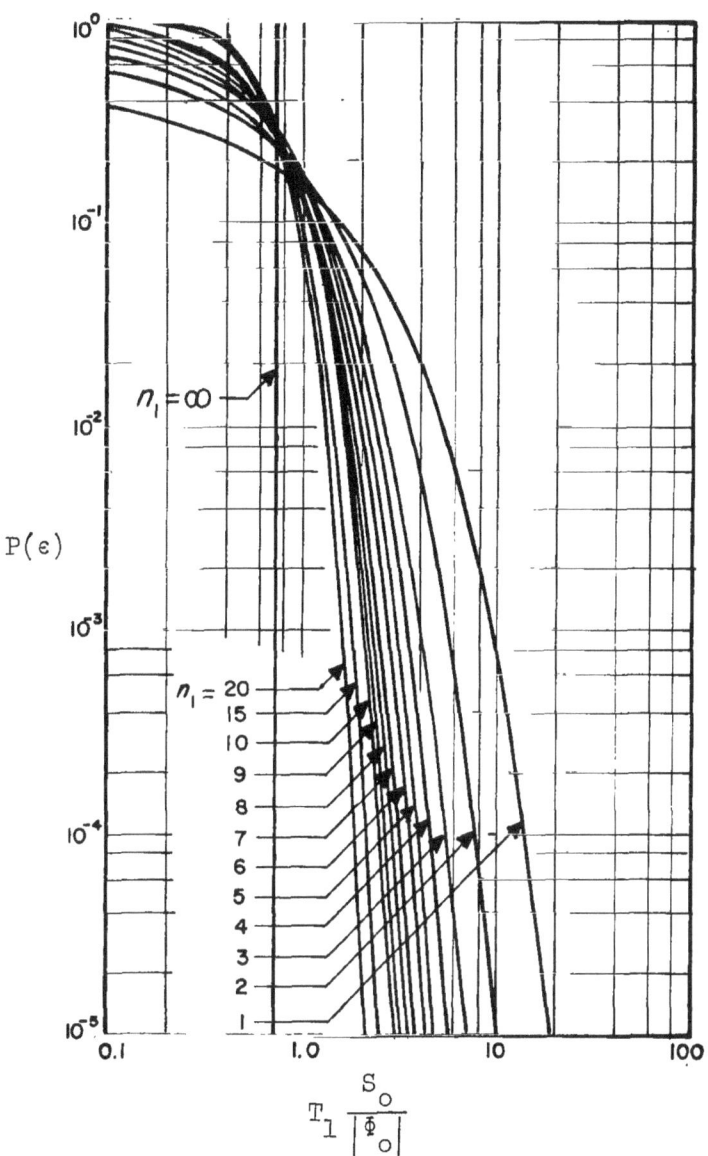

Figure F.1.5-1.- Error probability versus signal-to-noise density ratio

F-17

F.2 PCM Telemetry Subcarrier Demodulator

This section will develop an approximate analysis for the demodulator shown in figure F.2-1.

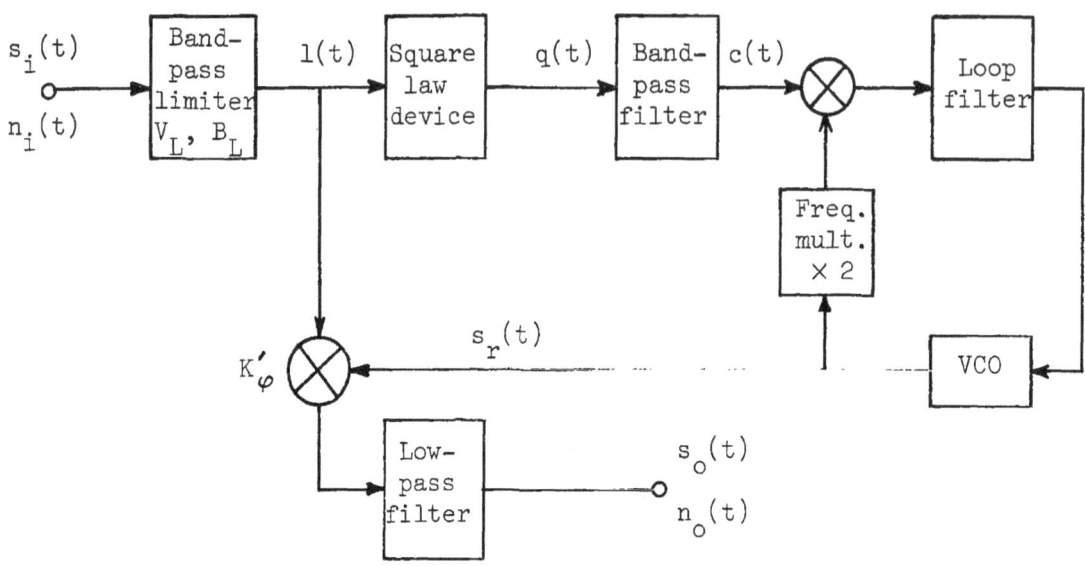

Figure F.2-1.- PCM telemetry subcarrier demodulator

The input signal is taken as purely phase modulated, of the form

$$s_i(t) = A \cos\left[\omega_c t + \varphi_s(t)\right] \qquad (1)$$

where the modulation function is biphase

$$\varphi_s(t) = \frac{\pi}{2} c_t(t) \qquad (2)$$

where

$c_t(t)$ = square waveform having only the values ± 1

The input noise is taken as a sample function of a narrow-band Gaussian process, band-limited to the limiter bandwidth, as

$$n_i(t) = x(t) \cos \omega_c t - y(t) \sin \omega_c t \qquad (3)$$

From equation B.3 (4), page B-6, the limiter output $l(t)$ is taken as

$$l(t) = 4 \frac{V_L}{\pi} \cos \left[\omega_c t + \psi(t) \right] \qquad (4)$$

This treatment will be limited to relatively high signal-to-noise ratios (SNR) in the limiter. For high limiter SNR,

$$l(t) \cong 4 \frac{V_L}{\pi} \cos \left[\omega_c t + \varphi_s(t) + \frac{y(t)}{A} \right] \qquad (5)$$

from equation B.2 (10), page B-4.

F.2.1 Output Data Treatment

The limiter signal is demodulated phase coherently by the product detector having gain constant K'_φ. It is assumed that the reference signal $s_r(t)$ from the VCO is essentially noiseless when the output data is usable. This implies that whenever the output data is usable, the subcarrier tracking loop is well above threshold. The reference signal is taken as

$$s_r(t) = - \sin \omega_c t \qquad (1)$$

The output signal and noise spectral density may be taken directly from equations B.4.2 (8) and (10), page B-9, as

$$s_o(t) = \frac{2}{\pi} V_L K'_\varphi \sin \varphi_s(t) \qquad (2)$$

$$\left| \Phi_{no} \right| = 8 \left[\frac{V_L K'_\varphi}{\pi A} \right]^2 \left| \Phi_{ni} \right| \qquad (3)$$

Using the identity of equation A.3 (3), page A-7, equation F.3 (2) may be substituted in (2) to give

$$s_o(t) = \frac{2}{\pi} V_L K'_\varphi c_t(t) \qquad (4)$$

A quantity which is useful for predicting data quality is the ratio of output data bit energy-to-noise spectral density $\frac{E}{|\Phi_{no}|}$. This is given as the bit rate R times the ratio of output power to noise spectral density.

$$\frac{E}{|\Phi_{no}|} = R \frac{S_o}{|\Phi_{no}|} = \frac{R \frac{4}{\pi^2} V_L^2 K_\varphi'^2}{8 \frac{V_L^2 K_\varphi'^2}{\pi^2 A^2} |\Phi_{ni}|} \tag{5}$$

$$\frac{E}{|\Phi_{no}|} = \frac{R}{2} \frac{A^2}{|\Phi_{ni}|} \tag{6}$$

$$\frac{E}{|\Phi_{no}|} = R \frac{S_i}{|\Phi_{ni}|} \tag{7}$$

where

$\frac{S_i}{|\Phi_{ni}|}$ = ratio of input subcarrier power to input noise spectral density

Equation (7) holds for reasonably high limiter SNR.

F.2.2 Reference Loop Treatment

The square law device squares the limiter signal $l(t)$ and passes all zonal energy near the second harmonic of the subcarrier frequency through the band-pass filter to the phase-locked loop. The output of the squaring device is

$$q(t) = l^2(t) = 8 \left[\frac{V_L}{\pi}\right]^2 \left\{1 + \cos\left[2\omega_c t + 2\psi(t)\right]\right\} \tag{1}$$

The driving signal for the loop is taken as the double frequency term,

$$c(t) = 8 \left[\frac{V_L}{\pi}\right]^2 \cos\left[2\omega_c t + 2\psi(t)\right] \qquad (2)$$

For reasonably high limiter SNR, equation (2) is well approximated by

$$c(t) = 8 \left[\frac{V_L}{\pi}\right]^2 \cos\left[2\omega_c t + 2\varphi_s(t) + 2\frac{y(t)}{A}\right] \qquad (3)$$

Due to the assumed square telemetry waveform, the signal term of equation (3) is identically

$$2\varphi_s(t) \equiv \pi\, c_t(t) \equiv \pm\pi \qquad (4)$$

then

$$c(t) = -8 \left[\frac{V_L}{\pi}\right]^2 \cos\left[2\omega_c t + 2\frac{y(t)}{A}\right] \qquad (5)$$

Comparison of equation (5) with equations B.2 (10) and B.2 (11), both on page B-4, shows that the phase noise spectral density for the PCM telemetry reference loop is given by

$$\left|\Phi_\varphi\right| = 8\frac{\left|\Phi_{ni}\right|}{A^2} \qquad (6)$$

Since $c(t)$ contains no signal modulation, the loop will not have modulation tracking error, except possibly for Doppler effects. Neglecting Doppler, the loop phase jitter is obtained from equation (6) and equation C.2.3 (1), page C-14, as

$$\sigma_\varphi^2 = 4 \left[\frac{N_i}{S_i}\right]_{B_N} \qquad (7)$$

The loop may be treated for threshold as in section C.2.4, employing equation (7) above. Equation (7) is valid for reasonably high limiter SNR.

F.3 The Residual Carrier Tracking Receiver (Ground)

The ground carrier tracking receiver, shown below in figure F.3-1, is a closed loop, phase tracking, double-superheterodyne receiver. This section will determine the equivalence between the receiver and a simple phase-locked loop as treated in appendix C.

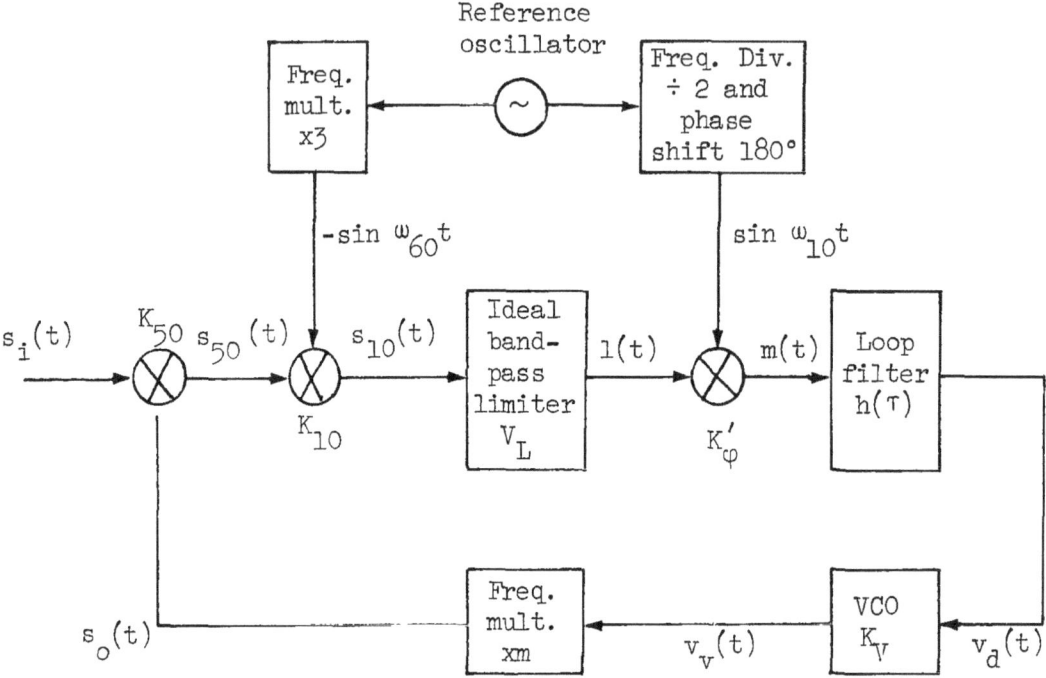

Figure F.3-1.- Carrier tracking receiver

In the figure, the various K's are amplitude transmission constants. The numbered subscripts refer to the nominal center frequencies of the various signals. That is, $s_{10}(t)$ represents a signal whose nominal center frequency is 10 megacycles. The frequency multiplication factor of the network between the VCO and the first mixer is m. The input signal is taken as in appendix C.

$$s_i(t) = A \cos \left[\omega_c t + \varphi_i(t) \right] \tag{1}$$

The VCO output voltage is taken as in appendix C, as

$$v_V(t) = -A_V \sin\left[\omega_V t + \varphi_V(t)\right] \tag{2}$$

where ω_V is not equal to ω_c. The first mixer injection signal is a frequency multiplied version of the VCO signal.

$$s_o(t) = -A_o \sin\left[m\omega_V t + m\varphi_V(t)\right] \tag{3}$$

The first intermediate frequency signal is one term of the product

$$K_{50} s_i(t) s_o(t) = -K_{50} A A_o \sin\left[m\omega_V t + m\varphi(t)\right] \cos\left[\omega_c t + \varphi_i(t)\right] \tag{4}$$

In particular,

$$s_{50}(t) = -\frac{K_{50} A A_o}{2} \sin\left[\omega_{50} t + \varphi_i(t) - m\varphi_V(t)\right] \; ; \; \omega_{50} \triangleq \omega_c - m\omega_V \tag{5}$$

The second intermediate frequency signal is one term of the product,

$$-K_{10} s_{50}(t) \sin \omega_{60} t = \frac{K_{10} K_{50} A A_o}{2} \sin\left[\omega_{50} t + \varphi_i(t) - m\varphi_V(t)\right] \sin \omega_{60} t \tag{6}$$

In particular,

$$s_{10}(t) = \frac{K_{10} K_{50} A A_o}{4} \cos\left[\omega_{10} t + m\varphi_V(t) - \varphi_i(t)\right] \; ; \; \omega_{10} \triangleq \omega_{60} - \omega_{50} \tag{7}$$

$s_{10}(t)$ is the limiter input signal. The limiter output signal $l(t)$ has an amplitude constant dependent only on limiting level V_L.

$$l(t) = \frac{4V_L}{\pi} \cos\left[\omega_{10} t + m\varphi_V(t) - \varphi_i(t)\right] \tag{8}$$

The multiplier signal $m(t)$ is taken as the low-frequency term, as in appendix C.

$$m(t) = -2K'_\varphi \frac{V_L}{\pi} \sin\left[m\varphi_V(t) - \varphi_i(t)\right] \qquad (9)$$

or

$$m(t) = 2K'_\varphi \frac{V_L}{\pi} \sin\left[\varphi_i(t) - m\varphi_V(t)\right] \qquad (10)$$

Equation (10) is identical in form to equation C.1 (5), page C-2. Therefore, the VCO output phase function $\varphi_V(t)$ may be written directly as

$$\varphi_V(t) = 2K'_\varphi \frac{V_L}{\pi} K_V \int_0^t \int_0^\infty h(\tau) \sin\left[\varphi_i(t-\tau) - m\varphi_V(t-\tau)\right] d\tau dt \qquad (11)$$

where

$h(\tau)$ = impulse response function of the loop filter

K_V = VCO constant

Multiplying both sides of equation (11) by m, we obtain

$$m\varphi_V(t) = 2K'_\varphi K_V m \frac{V_L}{\pi} \int_0^t \int_0^\infty h(\tau) \sin\left[\varphi_i(t-\tau) - m\varphi_V(t-\tau)\right] d\tau dt \qquad (12)$$

Equation (12) is identical in form to equation C.1 (9), page C-3. Therefore, by analogy, equation (12) describes a simple phase-locked loop with input phase function of $\varphi_i(t)$, output phase function of $m\varphi_V(t)$, and open loop gain (neglecting loop filter constant) of

$$K = \frac{2}{\pi} K'_\varphi K_V m V_L \qquad (13)$$

It is seen from equation (13) that the frequency multiplication constant m has been incorporated into the loop gain. The presence of the limiter in the loop may be expected to produce limiter effects, treated in section C.3.1, under conditions of low limiter SNR.

The conclusion to be drawn from the above is that the ground residual carrier tracking receiver may be treated for threshold as a normal phase-locked loop by the methods of section C.4.

F.4 The Residual Carrier Tracking Receiver (Spacecraft)

The spacecraft carrier tracking receiver, shown below in figure F.4-1, is a closed-loop, phase-tracking, double-superheterodyne receiver. This section will determine the equivalence between the receiver and a simple phase-locked loop as treated in appendix C.

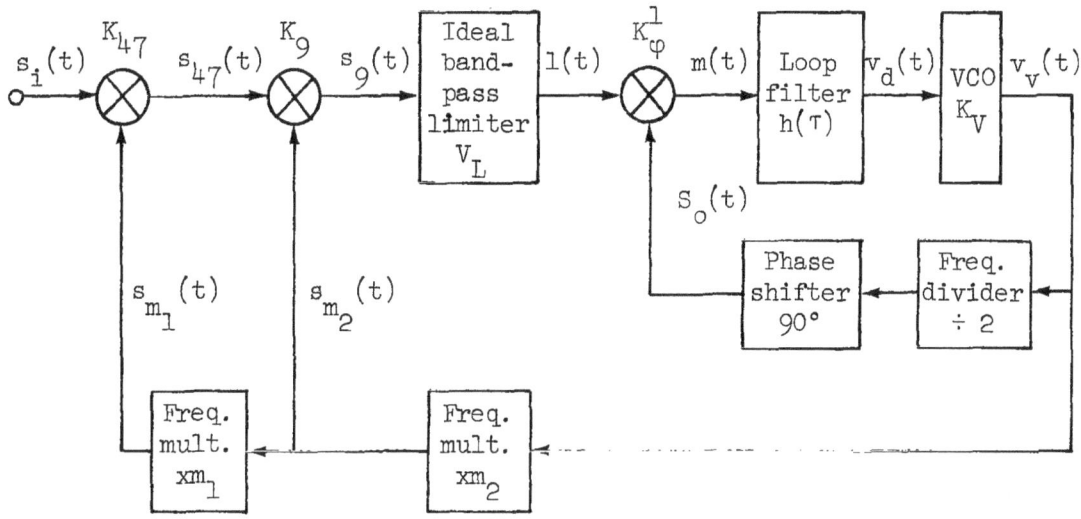

Figure F.4-1.- Carrier tracking receiver

In the figure, the various K's are amplitude transmission constants. The numbered subscripts refer to the nominal center frequencies of the various signals. That is, $s_{47}(t)$ represents a signal whose nominal center frequency is 47 megacycles. m_1 and m_2 are frequency multiplication factors for the networks between the VCO and the mixers.

The input signal is taken as in appendix C.

$$s_i(t) = A \cos \left[\omega_c t + \varphi_i(t) \right] \tag{1}$$

The VCO output signal is taken as

$$v_V(t) = A_V \cos\left[\omega_V t + \varphi_V(t)\right] \tag{2}$$

The VCO signal is frequency multiplied by a factor m_2 to obtain the second mixer injection signal $s_{m_2}(t)$.

$$s_{m_2}(t) = A_2 \cos\left[m_2\omega_V t + m_2\varphi_V(t)\right] \tag{3}$$

The second injection signal is frequency multiplied by a factor m_1 to obtain the first mixer injection signal $s_{m_1}(t)$.

$$s_{m_1}(t) = A_1 \cos\left[m_1 m_2 \omega_V t + m_1 m_2 \varphi_V(t)\right] \tag{4}$$

The first intermediate frequency signal $s_{47}(t)$ is one term of the product

$$K_{47} s_{m_1}(t) s_{m_2}(t) = K_{47} A A_1 \cos\left[\omega_c t + \varphi_i(t)\right] \cos\left[m_1 m_2 \omega_V t + m_1 m_2 \varphi_V(t)\right] \tag{5}$$

In particular

$$s_{47}(t) = \frac{K_{47} A A_1}{2} \cos\left[\omega_{47} t + \varphi_i(t) - m_1 m_2 \varphi_V(t)\right] \; ; \; \omega_{47} \triangleq \omega_c - m_1 m_2 \omega_V \tag{6}$$

The second intermediate frequency signal $s_9(t)$ is one term of the product

$$K_9 s_{m_2}(t) s_{47}(t) = \frac{K_9 K_{47} A A_1 A_2}{2} \cos\left[\omega_{47} t + \varphi_i(t) - m_1 m_2 \varphi_V(t)\right] \cdot$$

$$\cos\left[m_2 \omega_V t + m_2 \varphi_V(t)\right] \tag{7}$$

In particular,

$$s_9(t) = \frac{K_9 K_{47} A A_1 A_2}{4} \cos\left[\omega_9 t + \varphi_i(t) - m_2(1 + m_1)\varphi_V(t)\right] ;$$

$$\omega_9 \triangleq \omega_{47} - m_2 \omega_V \equiv \omega_c - m_2(1 + m_1)\omega_V \tag{8}$$

$s_9(t)$ is the limiter input signal. The limiter output signal $l(t)$ has an amplitude constant dependent only on limiting level V_L

$$l(t) = \frac{4V_L}{\pi} \cos\left[\omega_9 t + \varphi_i(t) - m_2(1 + m_1)\varphi_V(t)\right] \tag{9}$$

The other input to the phase detector is $s_o(t)$ which is the VCO signal, frequency divided by 2, and phase shifted by 90°.

$$s_o(t) = -A_o \sin\left[\omega_9 t + \frac{\varphi_V(t)}{2}\right] ; \quad \omega_9 \triangleq \frac{\omega_V}{2} \tag{10}$$

As in appendix C, the multiplier signal $m(t)$ is taken as the low-frequency term of the product of $l(t)$ and $s_o(t)$.

$$m(t) = \frac{2}{\pi} K'_\varphi V_L A_o \sin\left\{\varphi_i(t) - \left[\frac{2m_2(1 + m_1) + 1}{2}\right]\varphi_V(t)\right\} \tag{11}$$

Equation (11) is identical in form to equation C.1 (5), page C-2. Therefore, the VCO output phase function $\varphi_V(t)$ may be written directly as

$$\varphi_V(t) = \frac{2}{\pi} K'_\varphi V_L A_o K_V \int_o^t \int_o^\infty h(\tau) \sin\left\{\varphi_i(t - \tau)\right.$$

$$\left. - \left[\frac{2m_2(1 + m_1) + 1}{2}\right]\varphi_V(t - \tau)\right\} d\tau dt \tag{12}$$

where

$h(\tau)$ = impulse response function of the loop filter

K_V = VCO constant.

We next obtain

$$\left[\frac{2m_2(1 + m_1) + 1}{2}\right]\varphi_V(t) = \frac{2}{\pi} K'_\varphi V_L A_o K_V \left[\frac{2m_2(1 + m_1) + 1}{2}\right] \int_o^t \int_o^\infty$$

$$h(\tau) \sin\left\{\varphi_i(t - \tau) - \left[\frac{2m_2(1 + m_1) + 1}{2}\right]\varphi_V(t - \tau)\right\} d\tau dt$$

(13)

Equation (13) is identical in form to equation C.1 (9), page C-3. Therefore, by analogy, equation (13) describes a simple phase-locked loop with input phase function of $\varphi_i(t)$, output phase function of $\left[\frac{2m_2(1 + m_1) + 1}{2}\right]\varphi_V(t)$, and open loop gain (neglecting loop filter constant) of

$$K = \frac{2}{\pi} K'_\varphi V_L A_o K_V \left[\frac{2m_2(1 + m_1) + 1}{2}\right]$$

(14)

It is seen from equation (14) that the frequency multiplication constants m_1 and m_2 have been incorporated in the loop gain. Also, the presence of the limiter in the loop may be expected to produce limiter effects, treated in section C.3.1, under conditions of low limiter SNR.

It is interesting to note a difference between the ground carrier tracking receiver, treated in section F.3, and the spacecraft receiver. In the ground receiver, the input and output phase functions of the equivalent loop were the signal input phase function $\varphi_i(t)$ and the first mixer injection signal phase function $m\varphi_V(t)$. For conditions of lock, the first mixer injection signal tracked the input signal phase exactly, assuming no static phase error in the equivalent loop. For the spacecraft receiver, the input and output phase functions of the equivalent loop are the signal input phase $\varphi_i(t)$, and the function $\left[\frac{2m_2(1 + m_1) + 1}{2}\right]\varphi_V(t)$. This equivalent output phase function does not actually exist anywhere in the spacecraft receiver, as may be seen by examination of equations (1), (3), (4) and (10). However, for the case where the equivalent loop is locked with no loop error, that is, where

$$\varphi_V(t) = \left[\frac{2}{2m_2(1 + m_1) + 1}\right] \varphi_i(t) \tag{15}$$

it may be easily seen by substitution of equation (15) into equations (3), (4), (6), (9), and (10), that the internal phase-locked loop in the spacecraft receiver tracks exactly, as does the second mixer injection signal. The first mixer injection signal tracks the input signal phase somewhat in error.

The conclusion to be drawn from the above is that the spacecraft residual carrier tracking receiver may be treated for threshold as a normal phase-locked loop by the methods of section C.4.

F.5 The Spacecraft Turnaround Ranging Channel

The physical description and operation of the spacecraft channel which is used for "turnaround ranging" has been treated previously (ref. 13). This section will develop an approximate treatment for the channel. A rigorous treatment would be so complex as to be practically unusable.

The turnaround ranging channel, shown in figure F.5-1, comprises an ideal bandpass limiter, a coherent product detector, and a phase modulator. The up-link range code is demodulated, along with any subcarriers and noise in the bandpass of the limiter, and is remodulated along with subcarriers and noise onto the down-link.

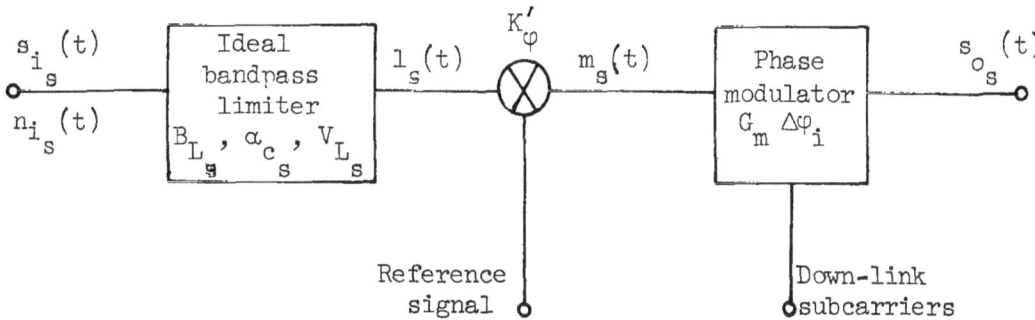

Figure F.5-1.- Spacecraft turnaround channel

The output signal of this channel, $s_{o_s}(t)$, differs from that assumed in the treatment of detection in appendices D and F due to several effects of the channel. The theory of appendices D and F must be slightly modified to account for these effects. The effects are an additional noise spectrum on the down-link due to turned-around noise, a suppression of the desired down-link signal due to remodulated noise, a suppression of the desired signal due to the spacecraft limiter, and a suppression of the desired down-link channels due to turned-around subcarriers.

The input signal to the limiter, $s_{i_s}(t)$, is taken in normal form as a sinusoid, phase modulated by the sum of range code plus an arbitrary number, L, of subcarriers.

$$s_{i_s}(t) = A_s \cos\left[\omega_c t + \varphi_s(t)\right] \tag{1}$$

where

$$\varphi_s(t) = \Delta\varphi_V c_r(t) + \sum_{j=1}^{L} \Delta\varphi_j \sin\left[\omega_j t + \varphi_j(t)\right] \tag{2}$$

The subscript j denotes up-link subcarriers explicitly.

The input noise is assumed to be a sample function of a Gaussian noise process, white, and band-limited to B_{L_s}, the physical bandwidth of the limiter.

$$n_{i_s}(t) = x_s(t) \cos \omega_c t - y_s(t) \sin \omega_c t \tag{3}$$

The subscript s denotes a spacecraft quantity.

From equation D.2 (10), page D-11, the multiplier signal driving the phase modulator is approximated as

$$m_s(t) \cong \frac{2}{\pi} K'_\varphi V_{L_s} \alpha_{L_s} \left[\sin \varphi_S(t) + \sqrt{\frac{4}{\pi}} \frac{y_s(t)}{A_s}\right] \tag{4}$$

where

α_{L_s} = limiter signal suppression constant

From equation D.2 (5), page D-10, α_{L_s} is approximated as

$$\alpha_{L_s}^2 = \frac{1}{1 + \frac{4}{\pi} \left[\frac{N_{i_s}}{S_{i_s}}\right]_{B_{L_s}}} \qquad (5)$$

where

$\left.\dfrac{N_{i_s}}{S_{i_s}}\right|_{B_{L_s}}$ = spacecraft limiter input noise-to-signal ratio

The phase modulator sums the ranging channel (signal) with the normal down-link subcarriers. A modulation gain G_r is applied to the ranging channel. The normal subcarriers have phase deviations $\Delta\varphi_i$. Then the output signal $s_{o_s}(t)$ is given as

$$s_{o_s}(t) = A_{o_s} \cos\left\{\omega_c t + \sum_{i=1}^{K} \Delta\varphi_i \sin\left[\omega_i t + \varphi_i(t)\right] + G_r m_s(t)\right\} \qquad (6)$$

For simplification, we may define a phase index $\Delta\varphi_m$ for the turnaround channel as

$$\Delta\varphi_m \triangleq \frac{2K'_\varphi V_{L_s} G_r}{\pi} \qquad (7)$$

then the phase modulator output is

$$s_{o_s}(t) = A_{o_s} \cos\left\{\omega_c t + \sum_{i=1}^{K} \Delta\varphi_i \sin\left[\omega_i t + \varphi_i(t)\right] \right.$$

$$\left. + \Delta\varphi_m \alpha_{L_s} \left[\sin\varphi_s(t) + \sqrt{\frac{4}{\pi}} \frac{y_s(t)}{A_s}\right]\right\} \qquad (8)$$

The input signal received at the ground is taken as

$$s_{i_g}(t) = A_g \cos \left\{ \omega_c t + \sum_{i=1}^{K} \Delta\varphi_i \sin \left[\omega_i t + \varphi_i(t)\right] \right.$$

$$\left. + \Delta\varphi_m \alpha_{L_s} \left[\sin \varphi_s(t) + \sqrt{\frac{4}{\pi}} \frac{y_s(t)}{A_s}\right] \right\} \tag{9}$$

where A_g differs from A_{o_s} according to the signal gains and attenuations between spacecraft and ground.

F.5.1 Equivalent Noise

Observation of equation (9) shows that the signal received at the ground contains phase noise of the form

$$\varphi_\varphi(t) = \sqrt{\frac{4}{\pi}} \frac{\Delta\varphi_m \alpha_{L_s} y_s(t)}{A_s} \tag{1}$$

The effects of this noise are two-fold. First, the noise appears at the ground as noise and lowers the effective ground input signal-to-noise ratio. Secondly, the phase noise removes power from the carrier and effectively suppresses the remaining modulation. The suppression effect has been treated by Middleton (ref. 14).

The assumption is made that the r.m.s. phase deviation of the signal due to the phase noise is small enough so that the resulting phase noise spectrum is not spread or broadened beyond that of the original spacecraft noise. Then an equivalent external "incremental" noise function $N_\Delta(t)$ may be postulated, which, when summed with an assumed "noiseless" received signal, gives the same phase noise as that of the actual received signal. That is, an equality is defined as

$$s_{i_g}(t) \triangleq A_g e^{-\frac{\sigma_{\varphi_s}^2}{2}} \cos \left\{ \omega_c t + \sum_{i=1}^{K} \Delta\varphi_i \sin \left[\omega_i t + \varphi_i(t)\right] \right.$$

$$\left. + \Delta\varphi_m \alpha_{L_s} \sin \varphi_s(t) \right\} + N_\Delta(t) \tag{2}$$

where

$$N_\Delta(t) = x_\Delta(t) \cos \omega_c t - y_\Delta(t) \sin \omega_c t \qquad (3)$$

and

$$\sigma^2_{\varphi_s} = \frac{4}{\pi} \frac{\Delta\varphi_m^2 \alpha_{L_s}^2}{A_s^2} y_s^2(t) \qquad (4)$$

$e^{-\sigma^2_{\varphi_s}}$ is the power suppression factor of Middleton (ref. 14) due to phase modulation by Gaussian noise having an r.m.s. value of σ_{φ_s}.

Assuming that the modulation bandwidth encompasses all the noise passed through the spacecraft limiter, equation (4) may be rewritten as

$$\sigma^2_{\varphi_s} = \frac{2}{\pi} \Delta\varphi_m^2 \alpha_{L_s}^2 \left[\frac{N_i}{S_i}\right]_{B_{L_s}} = \frac{1}{2} \left[\frac{\Delta\varphi_m^2}{1 + \frac{\pi}{4}\left[\frac{S_i}{N_i}\right]_{B_{L_s}}}\right] \qquad (5)$$

The incremental noise function may be related to the phase noise function as

$$\frac{Y_\Delta(t)}{A_g} = \sqrt{\frac{4}{\pi}} \frac{\Delta\varphi_m \alpha_{L_s} y_s(t)}{A_s} \qquad (6)$$

From equation (6) the noise spectral density of the incremental noise function may be inferred as

$$\left|\Phi_{n_\Delta}\right| = \frac{4}{\pi} \Delta\varphi_m^2 \alpha_{L_s}^2 \frac{A_G^2}{A_s^2} \left|\Phi_{n_{i_s}}\right| \qquad (7)$$

Dividing equation (7) by the value of the actual ground system noise spectral density gives

$$\frac{\left|\Phi_{n_\Delta}\right|}{\left|\Phi_{n_{i_g}}\right|} = \frac{4}{\pi} \Delta\varphi_m^2 \, \alpha_{L_s}^2 \, \frac{A_G^2}{A_s^2} \frac{\left|\Phi_{n_{i_s}}\right|}{\left|\Phi_{n_{i_g}}\right|} \tag{8}$$

then

$$\frac{\left|\Phi_{n_\Delta}\right|}{\left|\Phi_{n_{i_g}}\right|} = \frac{4}{\pi} \Delta\varphi_m^2 \, \alpha_{L_s}^2 \left[\frac{\left.\frac{S_{i_g}}{N_{i_g}}\right|_{B_{L_s}}}{\left.\frac{S_{i_s}}{N_{i_s}}\right|_{B_{L_s}}} \right] \tag{9}$$

and

$$\frac{\left|\Phi_{n_\Delta}\right|}{\left|\Phi_{n_{i_g}}\right|} = \frac{\Delta\varphi_m^2 \left[\left.\frac{S_{i_g}}{N_{i_g}}\right|_{B_{L_s}}\right]}{1 + \frac{\pi}{4}\left[\left.\frac{S_{i_s}}{N_{i_s}}\right|_{B_{L_s}}\right]} \tag{10}$$

It should be noted that the quantity $\left.\frac{S_{i_g}}{N_{i_g}}\right|_{B_{L_s}}$ represents a computation of the signal-to-noise ratio due to the actual ground system noise spectral density, taken in a bandwidth equal to the spacecraft limiter. If the normal thermal input noises to the spacecraft limiter and ground receiver are assumed to be uncorrelated, a total equivalent noise spectral density in the ground receiver may be postulated as

$$\left|\Phi_{n_T}\right| = \left|\Phi_{n_{i_g}}\right| + \left|\Phi_{n_\Delta}\right| \tag{11}$$

$$\left| \Phi_{n_T} \right| = \left| \Phi_{n_{i_g}} \right| \left[1 + \Delta\varphi_m^2 \frac{\left[\frac{S_{i_g}}{N_{i_g}} \right] B_{L_s}}{1 + \frac{\pi}{4} \left[\frac{S_{i_s}}{N_{i_s}} \right] B_{L_s}} \right] \quad (12)$$

Equation (12) is an approximation usable within a bandwidth narrower than the spacecraft limiter bandwidth and centered on the ground received carrier frequency.

F.5.2 Equivalent Signal

From equation F.5.1 (2), page F-32, the signal phase function of the turned-around channel as received at the ground is given as

$$\varphi_{s_T}(t) = \Delta\varphi_m \alpha_{L_s} \sin \varphi_s(t) \quad (1)$$

Employing the identities of appendix A, this signal is seen to be

$$\varphi_{s_T}(t) = \Delta\varphi_m \alpha_{L_s} \left(c_t(t) \sin(\Delta\varphi_r) \cos\left\{ \sum_{j=1}^{1} \Delta\varphi_j \sin\left[\omega_j t + \varphi_j(t) \right] \right\} \right.$$
$$\left. + \cos(\Delta\varphi_r) \sin\left\{ \sum_{j=1}^{1} \Delta\varphi_j \sin\left[\omega_j t + \varphi_j(t) \right] \right\} \right) \quad (2)$$

The turned-around signal will be approximated by only the primary code term and the first order subcarrier terms. Then

$$\varphi_{s_T}(t) \cong \Delta\varphi_m \alpha_{L_s} \left\{ \sin(\Delta\varphi_r) \prod_{j=1}^{L} J_0(\Delta\varphi_j) c_t(t) \right.$$

$$\left. + \sum_{j=1}^{L} 2 \cos(\Delta\varphi_r) J_1(\Delta\varphi_j) \prod_{\substack{h=1 \\ h \neq j}}^{L} \left[J_0(\Delta\varphi_h) \right] \sin\left[\omega_j t + \varphi_j(t) \right] \right\} \quad (3)$$

The equivalent noiseless signal received on the ground is expressed finally as

$$s_g(t) = A_g e^{\frac{-\sigma_{\varphi_s}^2}{2}} \cos\left\{ \omega_c t + \Delta\varphi_r \text{eff } c_t(t) + \sum_{j=1}^{L} \Delta\varphi_j \text{eff } \sin\left[\omega_j t + \varphi_j(t) \right] \right.$$

$$\left. + \sum_{i=1}^{K} \Delta\varphi_i \sin\left[\omega_i t + \varphi_i(t) \right] \right\} \quad (4)$$

where

$$\Delta\varphi_r \text{eff} = \Delta\varphi_m \alpha_{L_s} \sin(\Delta\varphi_r) \prod_{j=1}^{L} \left[J_0(\Delta\varphi_j) \right] \quad (5)$$

and

$$\Delta\varphi_j \text{eff} = 2\Delta\varphi_m \alpha_{L_s} \cos(\Delta\varphi_r) J_1(\Delta\varphi_j) \prod_{\substack{h=1 \\ h \neq j}}^{L} \left[J_0(\Delta\varphi_h) \right] \quad (6)$$

As in section A.3, the residual carrier term may be written as

$$s_{c_g}(t) = A_g e^{\frac{-\sigma_{\varphi_s}^2}{2}} \cos(\Delta\varphi_r \text{eff}) \prod_{j=1}^{L} \left[J_0(\Delta\varphi_j \text{eff}) \right] \prod_{i=1}^{K} \left[J_0(\Delta\varphi_i) \right] \cos \omega_c t$$

$$(7)$$

APPENDIX G

PHASE MODULATED SIGNAL DESIGN

This section treats the design and optimization of narrow deviation phase modulated sinusoidal carriers. The modulation functions are taken to be baseband functions and/or subcarriers. Only the determination of the various modulation indices is treated, since selection of subcarrier frequencies is a separable problem and is a function of the spectral extent of the modulated subcarriers themselves.

The signal to be treated is taken from equation A.3 (1), page A-6.

$$s(t) = A \cos \left\{ \omega_c t + \Delta\varphi_r c_t(t) + \sum_{i=1}^{K} \Delta\varphi_i \sin \left[\omega_i t + \varphi_i(t) \right] \right\} \quad (1)$$

where

A = carrier amplitude

ω_c = carrier radian frequency

$c_t(t)$ = ranging code, having only values ± 1

$\Delta\varphi_r$ = peak phase deviation of carrier by ranging code

$\Delta\varphi_i$ = peak phase deviation of carrier by the ith subcarrier

ω_i = radian frequency of the ith subcarrier

$\varphi_i(t)$ = effective phase modulation on the ith subcarrier

It is assumed that the restriction to small phase deviations insures that most of the signal power will be concentrated in the zero and first order signal products.

It is assumed that signal detection is performed using phase coherent product detectors, treated in appendix D, or a specialized ranging receiver, treated in section F.1.

There are two basic signal design criteria, subject to certain boundary conditions on the residual carrier and the amount of allowable intermodulation.

The first criterion is a set of minimum design goals for the information channels. These design goals are generally specified by a minimum channel signal-to-noise ratio and the bandwidth in which it is taken. The signal must be designed so that as carrier power is decreased in the presence of additive white Gaussian channel noise, the minimum design goals are met simultaneously.

The second criterion is that when the channel design goals are simultaneously achieved, the channel signal-to-noise ratios must be maximized within the capabilities of available carrier power.

A boundary condition is that satisfaction of the basic design criteria should not reduce the signal-to-noise ratio in the residual carrier channel below its minimum design goal.

A second boundary condition is that the amount of unusable power or intermodulation products resulting from the satisfaction of the two design criteria should not be overly large.

G.1 Solution for Modulation Indices

For a coherent product detector, the output signal-to-noise ratio for the jth modulated subcarrier, taken in a bandwidth B_{o_j}, is obtained from equation D.1.4.1 (6), page D-7, as

$$\left.\frac{S_{o_j}}{N_{o_j}}\right|_{B_{o_j}} = 2\cos^2(\Delta\varphi_r) J_1^2(\Delta\varphi_j) \prod_{\substack{i=1 \\ i \neq j}}^{K} \left[J_o^2(\Delta\varphi_i)\right] \left[\frac{S_i}{N_i}\right]_{B_{o_j}} \qquad (1)$$

where

$\left.\dfrac{S_i}{N_i}\right|_{B_{o_j}}$ = ratio of total signal power to input noise power in a bandwidth B_{o_j}

For either a coherent product detector or for the range clock receiver of section F.1, the output signal-to-noise ratio for the range code, taken in a bandwidth B_{o_m}, is taken from equation F.1.3 (6), page F-12, in the form

$$\left.\frac{S_{o_r}}{N_{o_r}}\right|_{B_{o_r}} = L_D L_K \sin^2(\Delta\varphi_r) \prod_{i=1}^{K} J_o^2(\Delta\varphi_i) \left[\frac{S_i}{N_i}\right]_{B_{o_r}} \qquad (2)$$

For a product detector

$$L_D = L_K = 1 \qquad (3)$$

In terms of the notation used in section F.1.3,

$$\frac{S_{o_r}}{N_{o_r}} = \frac{S_{c_1}}{N_{c_1}} \ ; \ B_{o_r} = B_N \qquad (4)$$

Equations (1) and (2) may be rearranged as

$$1 = \frac{2\cos^2(\Delta\varphi_r) J_1^2(\Delta\varphi_j) \prod_{\substack{i=1 \\ i \neq j}}^{K} \left[J_o^2(\Delta\varphi_i)\right] \frac{S_i}{|\Phi_{n_i}| 2B_{o_j}}}{\left.\frac{S_{o_j}}{N_{o_j}}\right|_{B_{o_j}}} \qquad (5)$$

$$1 = \frac{L_D L_K \sin^2(\Delta\varphi_r) \prod_{i=1}^{K} \left[J_o^2(\Delta\varphi_i)\right] \frac{S_i}{|\Phi_{n_i}| 2B_{o_r}}}{\left.\frac{S_{o_r}}{N_{o_r}}\right|_{B_{o_r}}} \qquad (6)$$

Equations (5) and (6) may be combined as

$$\frac{2\cos^2(\Delta\varphi_r)\, J_1^2(\Delta\varphi_j) \prod_{\substack{i=1 \\ i\neq j}}^{K}\left[J_o^2(\Delta\varphi_i)\right] \frac{S_i}{\left|\Phi_{n_i}\right| 2B_{o_j}}}{\left[\dfrac{S_{o_j}}{N_{o_j}}\right] B_{o_j}}$$

$$= \frac{L_D L_K \sin^2(\Delta\varphi_r) \prod_{i=1}^{K}\left[J_o^2(\Delta\varphi_i)\right] \frac{S_i}{\left|\Phi_{n_i}\right| 2B_{o_r}}}{\left[\dfrac{S_{o_r}}{N_{o_r}}\right] B_{o_r}}$$

(7)

Equation (7) may be simplified to

$$\frac{1}{B_{o_j}\left[\dfrac{S_{o_j}}{N_{o_j}}\right]B_{o_j}} \times \frac{J_1^2(\Delta\varphi_j)}{J_o^2(\Delta\varphi_j)} = \frac{L_D L_K}{2B_{o_r}\left[\dfrac{S_{o_r}}{N_{o_r}}\right]B_{o_r}} \times \tan^2(\Delta\varphi_r) \qquad (8)$$

In general, for range code plus K subcarriers, there are K equations of the form of equation (8). These may be written in summed form as

$$\sum_{i=1}^{K}\left\{\frac{1}{B_{o_i}\left[\dfrac{S_{o_i}}{N_{o_i}}\right]B_{o_i}} \times \frac{J_1^2(\Delta\varphi_i)}{J_o^2(\Delta\varphi_i)} = \frac{L_D L_K}{2B_{o_r}\left[\dfrac{S_{o_r}}{N_{o_r}}\right]B_{o_r}} \times \tan^2(\Delta\varphi_r)\right\} \qquad (9)$$

The bandwidths are treated as system constants, and the signal-to-noise ratios as independent variables. $\Delta\varphi_i$ and $\Delta\varphi_r$ are dependent variables. When the bandwidths and signal-to-noise ratios are assigned as minimum design goals, repeated simultaneous solution of the K sets of equations yields sets of solutions $\left(\Delta\varphi_r, \Delta\varphi_i\right)$ satisfying the first design criterion. The solutions are not unique as there are an infinite number of solutions.

For a signal having K subcarriers only, with range code deleted, the equations analagous to equations (9) are

$$\sum_{i=1}^{K-1} \left\{ \frac{1}{B_{o_i}\left[\frac{S_{o_i}}{N_{o_i}}\right]_{B_{o_i}}} \times \frac{J_1^2(\Delta\varphi_i)}{J_0^2(\Delta\varphi_i)} = \frac{1}{B_{o_{i+1}}\left[\frac{S_{o_{i+1}}}{N_{o_{i+1}}}\right]_{B_{o_{i+1}}}} \times \frac{J_1^2(\Delta\varphi_{i+1})}{J_0^2(\Delta\varphi_{i+1})} \right\} \tag{10}$$

For the special case of a subcarrier which is phase-shift keyed ±90°, it is easily seen that

$$B_{o_i}\left[\frac{S_{o_i}}{N_{o_i}}\right]_{B_{o_i}} \equiv R\frac{E}{|\Phi|} \tag{11}$$

where

   R = keying bit rate

   E = energy per bit

   $|\Phi|$ = value of channel noise spectral density

For the special case of a quadraphase subcarrier which is phase-shift keyed by two telemetry channels, it may be determined that

$$B_{o_i}\left[\frac{S_{o_i}}{N_{o_i}}\right]_{B_{o_i}} \equiv R_x \frac{E_x}{|\Phi|} + R_y \frac{E_y}{|\Phi|} \qquad (12)$$

where

R_x and R_y = telemetry bit rates

E_x and E_y = telemetry energies per bit

$|\Phi|$ = value of channel noise spectral density

For the special case of the turnaround ranging channel, a relation between the effective turned-around phase indices of the range code and up subcarriers on the down carrier may be obtained by dividing equations F.5.2 (6) by F.5.2 (5), page F-36. Then

$$\frac{\Delta\varphi_j \text{eff}}{\Delta\varphi_r \text{eff}} = \frac{2\Delta\varphi_m \alpha_{L_s} \cos(\Delta\varphi_r) J_1(\Delta\varphi_j) \prod_{\substack{h=1 \\ h\neq j}}^{L} J_0(\Delta\varphi_h)}{\Delta\varphi_m \alpha_{L_s} \sin(\Delta\varphi_r) \prod_{j=1}^{L} J_0(\Delta\varphi_j)} \qquad (13)$$

$$\frac{\Delta\varphi_j \text{eff}}{\Delta\varphi_r \text{eff}} = 2 \cot(\Delta\varphi_r) \frac{J_1(\Delta\varphi_j)}{J_0(\Delta\varphi_j)} \qquad (14)$$

where

$\Delta\varphi_j$ = deviation of the jth subcarrier on the up carrier

$\Delta\varphi_r$ = deviation of the range code on the up carrier

Substitution of equation (14) in equation (8) gives

$$B_{o_r}\left[\frac{S_{o_r}}{N_{o_r}}\right]_{B_{o_r}} \equiv B_{o_j}\left[\frac{S_{o_j}}{N_{o_j}}\right]_{B_{o_j}} \times 2\left[\frac{\Delta\varphi_r\text{eff}}{\Delta\varphi_j\text{eff}}\right]^2 \quad (15)$$

where, for the turnaround channel, correlation loss L_K and detection loss L_D are taken as identically unity.

G.2 Maximization of Subcarrier Channel Signal-to-noise Ratios

Inspection of equation G.1 (1), page G-2, shows that the signal-to-noise ratio in the jth subcarrier channel is proportional to a function of modulation indices given by

$$F_j(\Delta\varphi_r, \Delta\varphi_i) = \cos^2(\Delta\varphi_r) J_1^2(\Delta\varphi_j) \prod_{\substack{i=1 \\ i \neq j}}^{K} \left[J_o^2(\Delta\varphi_i)\right] \quad (1)$$

For a signal composed of K subcarriers and ranging code, maximization of all the subcarrier channel signal-to-noise ratios is obtained by maximizing equation (1) for any arbitrary j, using sets of $(\Delta\varphi_r, \Delta\varphi_i)$ which are solutions of equation G.1 (9), page G-4. Since simultaneous satisfaction of equation G.1 (9) sets all the subcarrier channel signal-to-noise ratios proportional to each other by constants, maximization of one subcarrier channel maximizes all subcarrier channels.

G.3 Boundary Condition on Residual Carrier

The signal-to-noise ratio for the residual carrier channel in its bandwidth B_c may be determined from equation A.3 (7), page A-8, to be

$$\left.\frac{S_c}{N_c}\right|_{B_c} = \cos^2(\Delta\varphi_r) \prod_{i=1}^{K} \left[J_o^2(\Delta\varphi_i)\right] \left[\frac{S_i}{N_i}\right]_{B_c} \quad (1)$$

Rearranging equation (1) and equating to equation G.1 (5), page G-3, we have

$$\frac{\cos^2(\Delta\varphi_r) \prod_{i=1}^{K} J_0^2(\Delta\varphi_i) \dfrac{S_i}{\left|\Phi_{n_i}\right| 2B_c}}{\left.\dfrac{S_c}{N_c}\right|_{B_c}}$$

$$= \frac{2\cos^2(\Delta\varphi_r) J_1^2(\Delta\varphi_j) \prod_{\substack{i=1 \\ i\neq j}}^{K} J_0^2(\Delta\varphi_i) \dfrac{S_i}{\left|\Phi_{n_i}\right| 2B_{o_j}}}{\left.\dfrac{S_{o_j}}{N_{o_j}}\right|_{B_{o_j}}}$$

(2)

or

$$\frac{1}{2B_c \left[\dfrac{S_c}{N_c}\right]_{B_c}} = \frac{1}{B_{o_j}\left[\dfrac{S_{o_j}}{N_{o_j}}\right]_{B_{o_j}}} \times \frac{J_1^2(\Delta\varphi_j)}{J_0^2(\Delta\varphi_j)} \qquad (3)$$

The boundary condition is obtained by stipulating that when the signal-to-noise ratio in the jth subcarrier channel falls to its design goal the signal-to-noise ratio in the residual carrier channel should be equal to or greater than its design goal. Then,

$$\frac{1}{B_{o_j}\left[\dfrac{S_{o_j}}{N_{o_j}}\right]_{B_{o_j}}} \times \frac{J_1^2(\Delta\varphi_j)}{J_0^2(\Delta\varphi_j)} \leq \frac{1}{2B_c\left[\dfrac{S_c}{N_c}\right]_{B_c}} \qquad (4)$$

G.4 Signal Efficiency

From equations A.3 (6) and (7), page A-8, it may be determined that the powers residing in the residual carrier component, prime code component, and first order subcarrier components of the modulated signal are given, respectively, by

$$P_{carrier} = \left\{ \cos^2(\Delta\varphi_r) \prod_{i=1}^{K} J_o^2(\Delta\varphi_i) \right\} \frac{A^2}{2} \qquad (1)$$

$$P_{code} = \left\{ \sin^2(\Delta\varphi_r) \prod_{i=1}^{K} J_o^2(\Delta\varphi_i) \right\} \frac{A^2}{2} \qquad (2)$$

$$P_{subcarriers} = \left\{ 2\cos^2(\Delta\varphi_r) \sum_{j=1}^{K} \left[J_1^2(\Delta\varphi_j) \prod_{\substack{i=1 \\ i \neq j}}^{K} J_o^2(\Delta\varphi_i) \right] \right\} \frac{A^2}{2} \qquad (3)$$

The total usable power, or effective power, is given as the sum of the three component powers as

$$P_{eff} = \left[\prod_{i=1}^{K} J_o^2(\Delta\varphi_i) \right] \left[1 + 2\cos^2(\Delta\varphi_r) \sum_{i=1}^{K} \frac{J_1^2(\Delta\varphi_i)}{J_o^2(\Delta\varphi_i)} \right] \frac{A^2}{2} \qquad (4)$$

The percent, or decimal, effective power, referred to the total carrier power is

$$\% P_{eff} = \left[\prod_{i=1}^{K} J_o^2(\Delta\varphi_i) \right] \left[1 + 2\cos^2(\Delta\varphi_r) \sum_{i=1}^{K} \frac{J_1^2(\Delta\varphi_i)}{J_o^2(\Delta\varphi_i)} \right] \qquad (5)$$

The difference between effective power and total power is unusable power composed of other modulation products. $\% P_{eff}$ may be used as a boundary condition on channel maximization.

APPENDIX H

SUPPLEMENTARY THEORY

H.1 The Equivalent Noise Bandwidth of Linear Networks

Figure H.1-1 shows a diagram of a two-port linear network which is described by an input-output voltage transfer function $G(s)$ in the complex variable s.

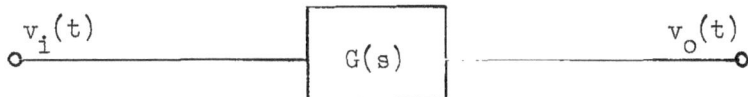

Figure H.1-1.- Linear network model

The transfer function is defined as

$$G(s) = \frac{V_o(s)}{V_i(s)} \qquad (1)$$

where

$V_o(s)$ = unilateral Laplace transform of the output voltage function $v_i(t)$

$V_i(s)$ = unilateral Laplace transform of the input voltage function $v_o(t)$

Since the network is linear, the principle of superposition applies and the treatment of the network for noise inputs may be made independently of considerations about the presence of signal.

The input to the network is taken as $n_i(t)$, a sample function of a Gaussian process. $n_i(t)$ is a function with finite non-zero power. That is,

$$0 < \overline{n_i^2(t)} < \infty \qquad (2)$$

where the bar denotes "average." It is assumed that the input noise has some noise spectral density, $\Phi_{n_i}(j\omega)$, which is a real, even function of the imaginary variable $j\omega$.

Papoulis (ref. 15) has shown that for finite power inputs to linear systems, the output spectral density may be written as

$$\Phi_{n_o}(j\omega) = \Phi_{n_i}(j\omega) |G(j\omega)|^2 \tag{3}$$

where

$$|G(j\omega)|^2 = \lim_{s \to j\omega} \left[G(s)G(-s)\right] = G(j\omega)G(-j\omega) \tag{4}$$

The total noise power out of the linear network is obtained by integrating the output spectral density with respect to frequency.

$$N_o = \frac{1}{2\pi j} \int_{-j\infty}^{j\infty} \Phi_{n_o}(j\omega) dj\omega = \frac{1}{2\pi j} \int_{-j\infty}^{j\infty} \Phi_{n_i}(j\omega) |G(j\omega)|^2 dj\omega \tag{5}$$

For the case where the input spectral density is constant or flat with value $\left|\Phi_{n_i}\right|$, in the regions of $G(j\omega)$ of interest, the integral may be rewritten as

$$N_o = \frac{1}{2\pi j} \left|\Phi_{n_i}\right| \int_{-j\infty}^{j\infty} |G(j\omega)|^2 dj\omega \tag{6}$$

Note that the use of transfer functions which exist for positive and negative frequencies implies the use of input spectral densities which are also "two-sided." This is no cause for alarm and is merely a consequence of the use of Fourier transforms. The Fourier transform of a real-time function is always two-sided. If, in the physical world where only positive frequencies have meaning, a "real" one-sided spectral density N is given as

$$N = KT \tag{7}$$

where

K = Boltzmann's constant

T = temperature

then the equivalent "two-sided" spectral density is simply

$$\left| \Phi_n \right| = \frac{N}{2} = \frac{KT}{2} \tag{8}$$

It is possible to define an equivalent "square" transfer function having constant amplitude G_r, some reference amplitude of the original transfer function, and a transmission bandwidth (two-sided) of $j2\Delta\omega_N$. This equivalent transfer function is defined such that the power transmitted through it from a white, Gaussian input density is exactly equal to the power transmitted through the original transfer function. The equivalence is made by equating output noise powers

$$N_o = \frac{1}{2\pi j} \left| \Phi_{n_i} \right| \int_{-j\infty}^{j\infty} \left| G(j\omega) \right|^2 dj\omega = \frac{1}{2\pi j} G_r^2 \left| \Phi_{n_i} \right| j2\Delta\omega_N \tag{9}$$

then

$$j2\Delta\omega_N = \frac{1}{G_r^2} \int_{-j\infty}^{j\infty} \left| G(j\omega) \right|^2 dj\omega \tag{10}$$

The treatment will now be confined to transfer functions which are ratios of polynominals in s, of the form

$$G(s) = \frac{P(s)}{Q(s)} \tag{11}$$

where the degree of $Q(s)$ is at least one greater than $P(s)$ and all the coefficients of s are real.

The left-hand quantity of equation (10) is called the "imaginary two-sided equivalent noise radian bandwidth." The real one-sided equivalent noise bandwidth in cycles is related as

$$B_N = \frac{\Delta\omega_N}{2\pi} \tag{12}$$

B_N is the physical square bandwidth through which equal power will be transmitted as that transmitted through the related transfer function. For transfer functions of the type specified, contour integration and the theory of residues may be used to solve equation (10).

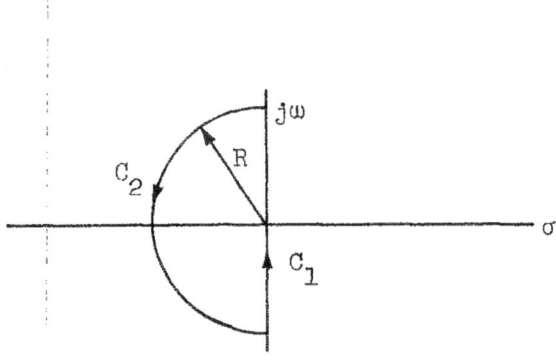

Figure H.1-2.- Contour of integration

Along the contour shown in the above figure, the equality holds.

$$\lim_{|R|\to\infty} \oint_{C_1,C_2} G(s)G(-s)ds = \lim_{|R|\to\infty} \int_{-j|R|}^{j|R|} G(j\omega)G(-j\omega)dj\omega$$

$$+ \lim_{|R|\to\infty} \int_{C_2} G(s)G(-s)ds \tag{13}$$

For large $|R|$ the transfer functions treated are of the order of $1/s$ and the integrand $G(s)G(-s)$ is of the order $1/s^2$. For these transfer functions and large $|R|$ the integral along path C_2 approaches zero.

then, by the theory of residues

$$\int_{-j\infty}^{j\infty} G(j\omega)G(-j\omega)dj\omega = 2\pi j \sum \left\{ \begin{array}{l} \text{Residues of } G(s)G(-s) \\ \text{in the left-half plane} \end{array} \right\} \quad (14)$$

and

$$2\Delta\omega_N = \frac{2\pi}{G_r^2} \sum \left\{ \begin{array}{l} \text{Residues of } G(s)G(-s) \\ \text{in the left-half plane} \end{array} \right\} \quad (15)$$

or

$$B_N = \frac{1}{2G_r^2} \sum \left\{ \begin{array}{l} \text{Residues of } G(s)G(-s) \\ \text{in the left-half plane} \end{array} \right\} \quad (16)$$

H.2 Equivalent Noise Temperature of Linear Systems

H.2.1 Single Networks

Every linear network, active or passive, contributes noise to a signal passing through it. For purposes of prediction, it is important that the noise properties of the networks dealt with be known.

It is well known that a resistance having a physical temperature T_R produces a white Gaussian one-sided noise spectral density (available power).

$$\Phi_R(f) = KT_R \frac{\text{joules}}{\text{cycle per second}} \quad (1)$$

where

K = Boltzmann's constant

It is possible to attribute noise produced by a linear network to an imaginary resistance at the network input, matched to the input, and to consider the network itself noiseless. The temperature of this imaginary resistance which would be required to produce the network noise if the network were noiseless is called the network's "equivalent noise temperature." Since the network is assumed linear, the presence of a signal or other uncorrelated noise does not influence the network self-generated noise or its equivalent noise temperature.

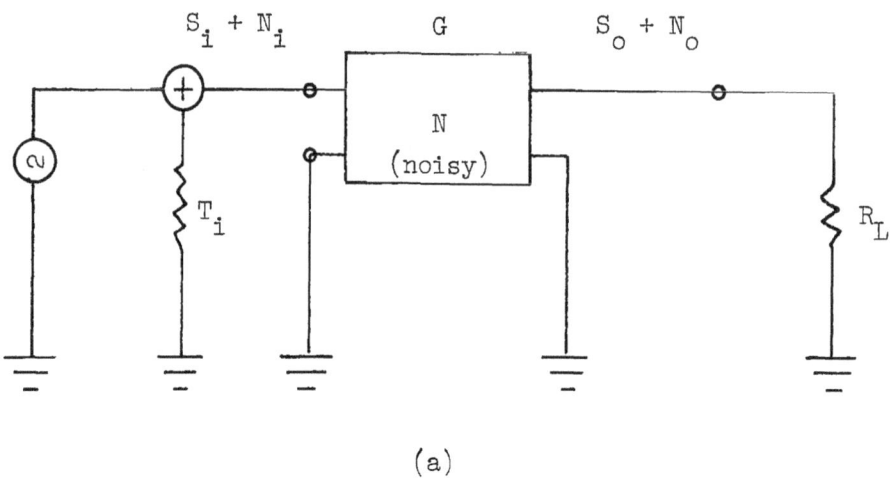

(a)

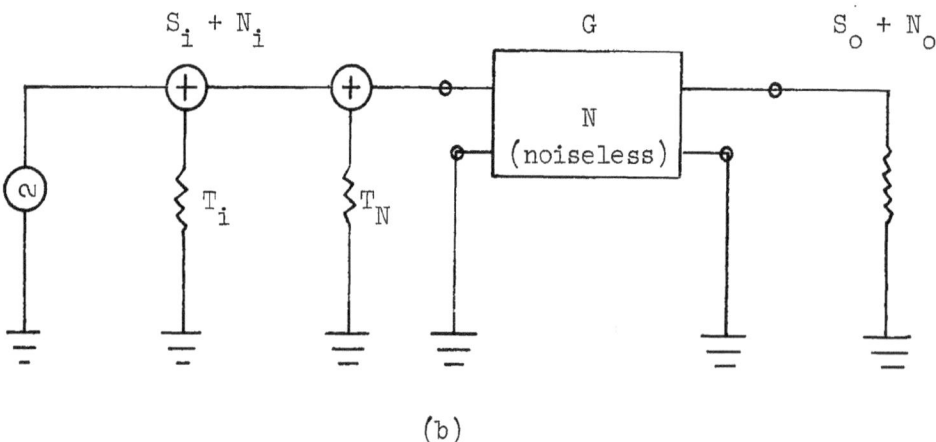

(b)

Figure H.2.1-1.- Equivalent noise temperature of a noisy linear network

Figure H.2.1-1 shows the resolution of a linear noisy network N having power gain G into a noiseless network fed by a resistor having equivalent noise temperature T_N. The summing junctions are a conceptual aid, indicating the summing of temperatures. They are not physical circuit summing points.

H-6

In figure H.2.1-1 (a),

$$\left.\begin{aligned} S_o &= GS_i \\ N_o &= N_N + G_{N_i} = N_N + GKT_i \text{ (per cycle of equivalent noise bandwidth)} \end{aligned}\right\} \quad (2)$$

where

N_N = noise contributed by the noisy networks

In figure H.2.1-1 (b),

$$\left.\begin{aligned} S_o &= GS_i \\ N_o &= N_N + GN_i - GKT_N + GKT_i \\ N_o &= GK\left(T_N + T_i\right) \end{aligned}\right\} \quad (3)$$

It is seen that in terms of equivalent noise temperature and on a per cycle equivalent noise basis, the ratio of input to output signal-to-noise ratios (noise figure) is given as:

$$\frac{\frac{S_i}{N_i}}{\frac{S_o}{N_o}} = \frac{\frac{S_i}{KT_i}}{\frac{GS_i}{GK(T_N + T_i)}} = 1 + \frac{T_N}{T_i} \quad (4)$$

This ratio is a figure of merit, which, when equal to one, indicates a noiseless system.

H.2.2 Cascaded Networks

There are generally two situations where it is desirable to obtain an equivalent noise temperature for two or more networks in cascade. The first case is for two noisy networks each having power gains greater than unity. It is desirable to have an equivalent temperature referenced to the input of the first network. The second case is for a noisy network with power gain greater than unity, fed by a passive network with power gain less than unity, fed by some "input" noise temperature. It is desirable to obtain an equivalent noise temperature referenced to the input of the network with gain greater than unity.

Consider two arbitrary networks in cascade, having power gains G1 and G2 and equivalent noise temperatures, referred to the individual network inputs of T_{1_e} and T_{2_e}, respectively.

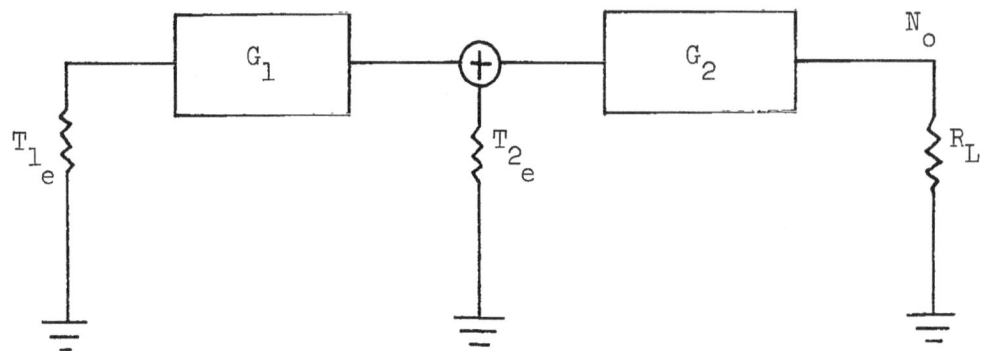

Figure H.2.2-1.- Cascaded linear noisy networks

The available output noise power is

$$N_o = G_2 K T_{2_e} + G_2 G_1 K T_{1_e} \qquad (1)$$

$$N_o = K G_2 G_1 \left[\frac{T_{2_e}}{G_1} + T_{1_e} \right] \qquad (2)$$

$$N_o = K \left[G_2 T_{2_e} + G_1 G_2 T_{1_e} \right] \qquad (3)$$

Equation (2) shows that the equivalent noise temperature referred to the input of the cascaded networks is

$$T_{n_i} = T_{1_e} + \frac{T_{2_e}}{G_1} \qquad (4)$$

Equation (3) shows that the equivalent temperature referred to the output of the cascaded networks is

$$T_{n_o} = G_2 \left[T_{2_e} + G_1 T_{1_e} \right] \qquad (5)$$

It can be shown (ref. 16) that the equivalent input noise temperature of a linear, bilateral, passive network, having power gain G_p, whose physical temperature is T_p, is given as

$$T_{p_e} = \left[\frac{1}{G_p} - 1\right] T_p \tag{6}$$

Case 1: For the special case of two noisy networks with power gains greater than unity, equation (4) shows that the equivalent noise input temperature is

$$T_{n_i} = T_{1_e} + \frac{T_{2_e}}{G_1} \tag{7}$$

For G_1 sufficiently high, T_1 denotes the expression.

Case 2: For the special case of a noisy hi-gain network, fed by a lossy passive network, fed by some input temperature T_i, figure H.2.2-2 applies.

$$T_e = T_{2_e} + G_L T_i + \left[1 - G_L\right] T_p \tag{8}$$

If the lossy network is defined by its attenuation or loss factor L where

$$1 < L = \frac{1}{G_L} < \infty \tag{9}$$

then

$$T_e = T_{2_e} + \frac{T_i}{L} + \left[1 - \frac{1}{L}\right] T_p \tag{10}$$

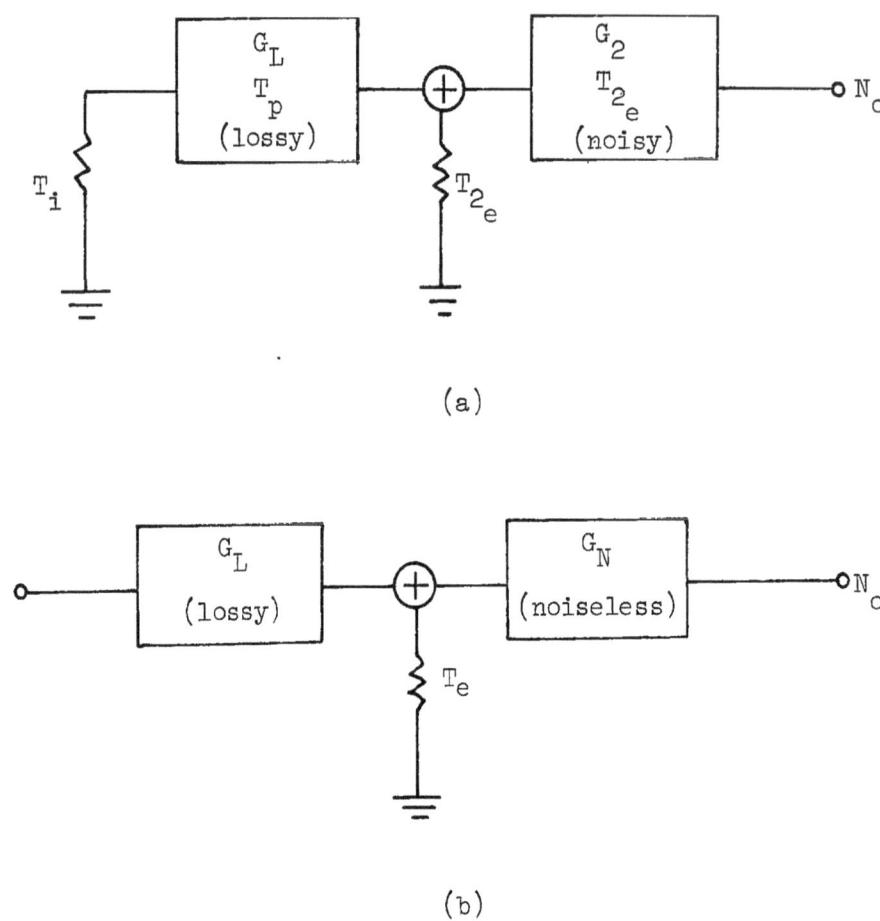

Figure H.2.2-2.- Cascaded passive and noisy networks

H.3 The Band-pass Amplitude Limiter

This section sets down, in the notation used elsewhere in this paper, certain pertinent results of the classic analysis of Davenport (ref. 4). The analysis was performed for an ideal "snap-action" limiter followed by an ideal band-pass filter. The driving signal was taken as a constant amplitude sinusoid embedded in narrow-band Gaussian noise. Narrow-banding of the input noise is taken to imply band-pass filtering at the input to the limiter. The model is shown in figure H.3-1. The output spectrum of the limiter itself contains spectral contributions centered not only at the fundamental input center frequency, but also at harmonics of the input center frequency. The action of the output band-pass filter is to allow transmission of only the energy centered on the

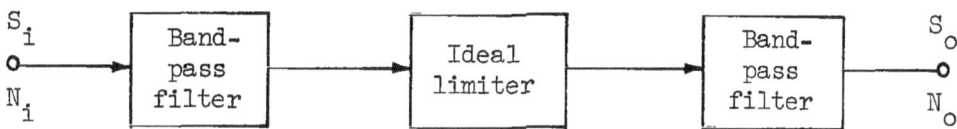

Figure H.3-1.- Band-pass limiter model

fundamental input frequency. For an output filter sufficiently wide to pass all the zonal energy centered on the fundamental frequency, Davenport's analysis shows that the output signal and output noise are related to the input signal-to-noise ratio (SNR) by rather complicated expressions involving the confluent hypergeometric function (ref. 17). The important results are reproduced below. Figure H.3-2 is a graph of output noise power and output signal power versus input SNR.

Observation of figure H.3-2 shows that the total output power of the band-pass filter is constant and is given by

$$P_o = S_o + N_o = 8\left[\frac{V_L}{\pi}\right]^2 \tag{1}$$

where

V_L = voltage limiting level

At low input SNR it is evident from figure H.3-2 that the output signal power is suppressed from the value at high input SNR. This suppression may be expressed through use of a signal voltage suppression factor α_L such that

$$S_o = \alpha_L^2 P_o \tag{2}$$

Martin (ref. 7) has approximated α_L by

$$\alpha_L^2 = \frac{1}{1 + \frac{4}{\pi}\left[\frac{N_i}{S_i}\right]} \tag{3}$$

where $\frac{N_i}{S_i}$ is the input noise-to-signal ratio. The actual and approximate α_L^2 are plotted in figure H.3-3.

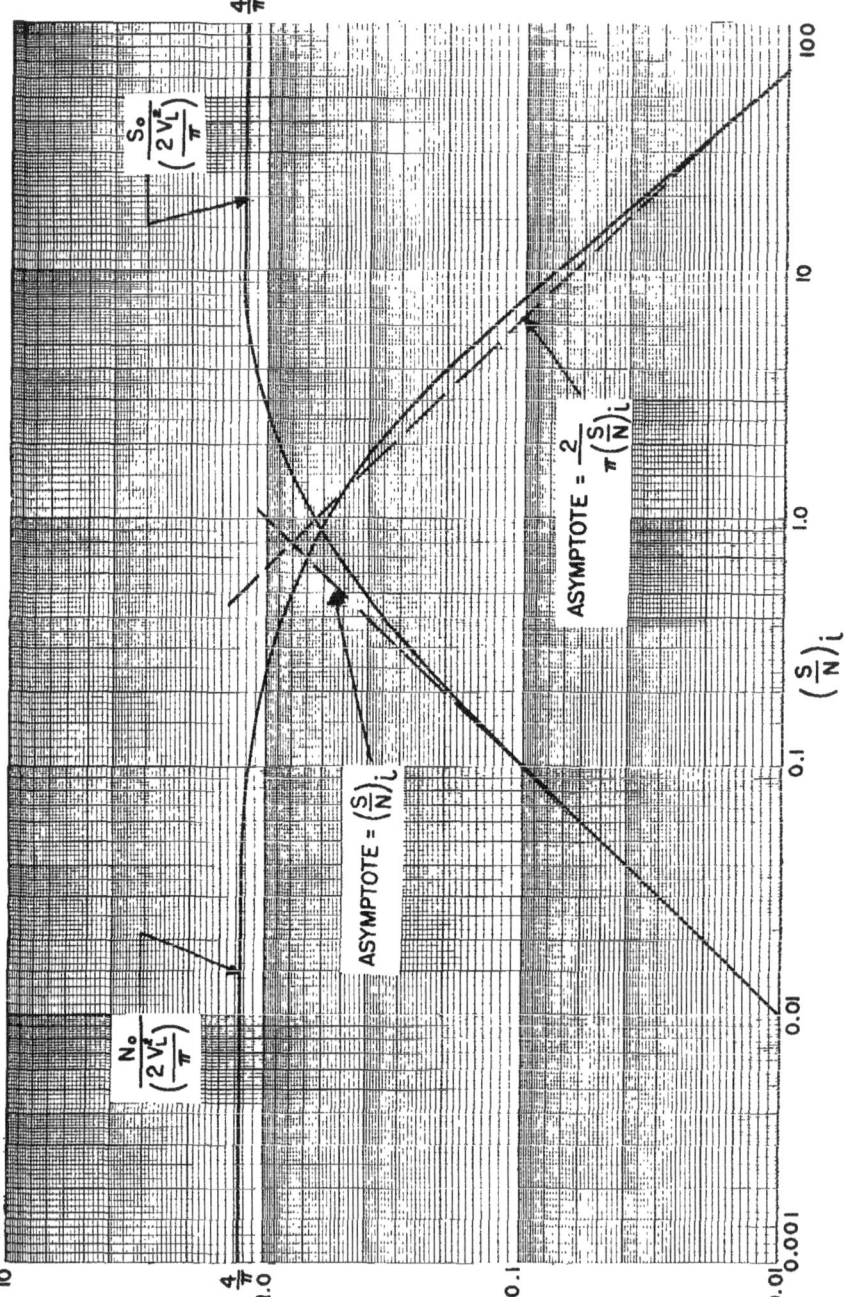

Figure H.3-2.- Limiter signal and noise suppression versus input SNR

H-12

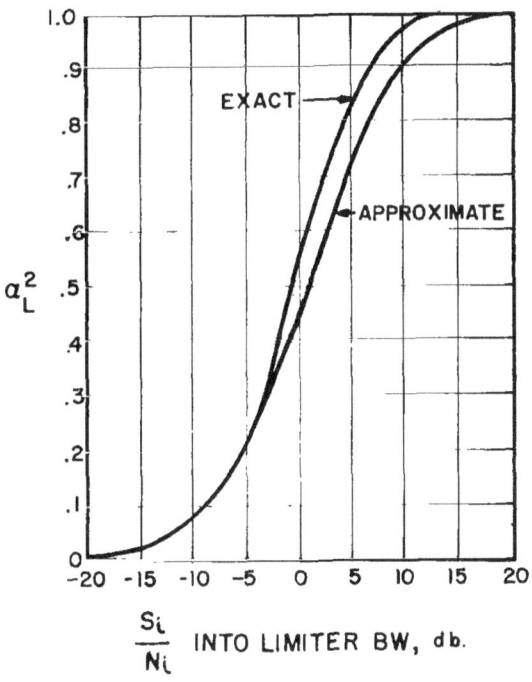

Figure H.3-3.- Exact and approximate signal suppression

Davenport's analysis was for the case of an unmodulated sinusoidal signal. For the purpose of simplifying the analysis in this paper, the assumption will be made that the results cited above apply equally to angle modulated sinusoidal signals.

H.4 The Range Equation

This section will derive the signal-to-noise ratio, computed in an arbitrary bandwidth B at some reference point of a radio receiver, due to transmission of radio energy from a transmitter which is physically separated from the receiver by a distance R.

Figure H.4-1 shows the model of the communications link to be used in the derivation. A transmitter with output power P_T feeds an antenna with power gain G_T through a lossy network having power gain G_{LT}. G_{LT} is a number less then unity. The energy from the transmitting antenna propagates across the distance R to the receiving antenna which has a power gain G_R. Only a portion of the transmitted energy

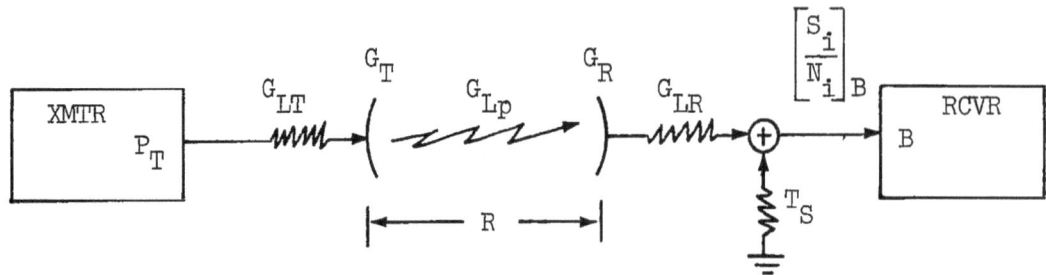

Figure H.4-1.- Communication link model

is intercepted by the receiving antenna. This loss of power is attributed to a propagation power gain G_{Lp} less than unity. The power intercepted by the antenna is passed through a lossy network, having power gain G_{LR}, to the reference point of the receiver. Noise in the link is attributed to an external system noise temperature T_S summed with the signal at the reference point.

The reference point for determining the signal-to-noise ratio is generally the input of a stage which has sufficient power gain so that addition of noise by subsequent stages is negligible.

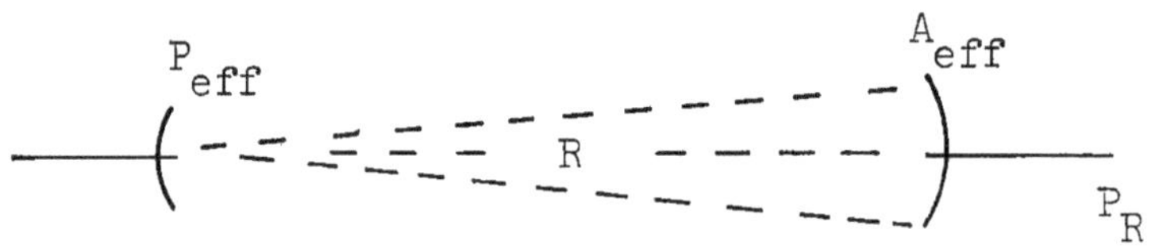

Figure H.4-2.- Antenna geometry

Figure H.4-2 details the geometry of the antenna system. The transmitting antenna gain is due to a bunching of the transmitted energy into a beam. This beam effect then raises the area power density of the receiving antenna. The effect is the same as if a higher effective power P_{eff} had been radiated by the transmitter.

$$P_{eff} = G_{LT} G_T P_T \tag{1}$$

The receiving antenna has an effective radio frequency area A_{eff} which is the area of the passing radio wave from which the antenna extracts all energy.

The area power density at the receiving antenna is given by simple geometry as

$$\eta = \frac{G_{Lp} P_{eff}}{4\pi R^2} \tag{2}$$

where

G_{Lp} = effect of propagation loss

The amount of power extracted by the receiving antenna is

$$P_R = \eta A_{eff} = \frac{G_{Lp} P_{eff} A_{eff}}{4\pi R^2} \tag{3}$$

$$P_R = \frac{G_{Lp} G_{LT} G_T P_T A_{eff}}{4\pi R^2}$$

The effective area of an antenna for receiving has been related to the power gain for transmitting by Friis (ref. 18) and others as

$$A_{eff} = \frac{G_R \lambda^2}{4\pi} \tag{5}$$

where

λ = wave length of the radio energy

then

$$P_R = G_{Lp} G_{LT} G_T G_R \left[\frac{\lambda}{4\pi R}\right]^2 P_T \tag{6}$$

The power reaching the receiver reference point is

$$S_i = G_{Lp} G_{LT} G_{LR} G_T G_R \left[\frac{\lambda}{4\pi R}\right]^2 P_T \tag{7}$$

The loss of power due to geometry may be attributed to a "space loss" having power gain G_{Ls}.

$$G_{Ls} = \left[\frac{\lambda}{4\pi R}\right]^2 = \left[\frac{C}{4\pi f R}\right]^2 \tag{8}$$

where

C = velocity of light

f = radio energy frequency

then

$$S_i = G_{Lp} G_{Ls} G_{LT} G_{LR} G_T G_R P_T \tag{9}$$

The effect of K different power gains may be expressed compactly as

$$S_i = P_T \prod_{i=1}^{K} G_i \tag{10}$$

The power gains less than unity may be expressed as

$$G_L = \frac{1}{L} \tag{11}$$

then equation (10) may be written

$$S_i = P_T \frac{\prod_{i=1}^{K} G_i}{\prod_{j=1}^{L} L_j} \tag{12}$$

The noise behavior of the receiver may be attributed to an input two-sided noise spectral density $\Phi_{n_i}(f)$ where

$$\Phi_{n_i}(f) = \frac{KT_s}{2} = \left|\Phi_{n_i}\right| \tag{13}$$

H-16

over some frequency range of interest where K is Boltzman's constant and T_s is the equivalent system noise temperature. The noise in an arbitrary bandwidth B is given as

$$N_i\big|_B = |\Phi_{n_i}| 2B \qquad (14)$$

or

$$N_i\big|_B = KT_s B \qquad (15)$$

The signal-to-noise ratio at the reference point is now written

$$\frac{S_i}{N_i}\bigg|_B = \frac{P_T}{KT_s B} \times \frac{\prod_{i=1}^{K} G_i}{\prod_{j=1}^{L} L_j} = \frac{P_T G_{LP} G_{LT} G_{LR} G_T G_R}{KT_s B} \left[\frac{c}{4\pi f R}\right]^2 \qquad (16)$$

H.5 Antenna Polarization Loss

The polarization of a radio wave is defined according to the space orientation of its electric vector. For an Earth bound receiving station, a radio wave whose electric vector is perpendicular to the Earth's surface is said to be vertically polarized; a wave whose electric vector is parallel to the Earth's surface is horizontally polarized. Both are special cases of linear polarization, wherein the electric vector always lies in one unique plane which is perpendicular to the plane of the wave. A more general type of polarization is elliptic polarization where the electric vector rotates in the plane of the wave and varies in amplitude as a function of rotation angle. Figure H.5-1 applies.

Figure H.5-1 shows a plane wave propagating out of the page, whose plane lies in the page. The sense of rotation of the electric vector with respect to the direction of propagation is clockwise. This is defined as "right-hand" polarization. The locus of electric vector amplitude is an ellipse, hence the name elliptic polarization. The ratio of minor axis length to major axis length for the ellipse is called the "axial ratio" or "ellipticity ratio". It is seen that linear polarization is a special case of elliptical polarization where the axial ratio is zero. The special case for axial ratio of one is called circular polarization.

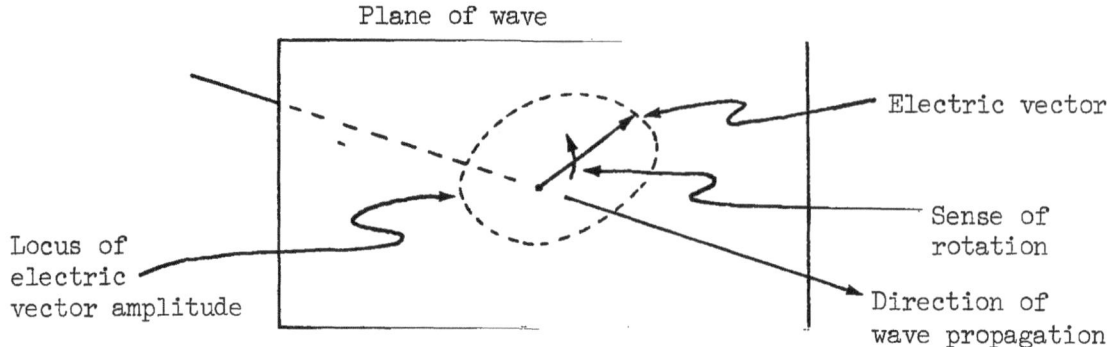

Figure H.5-1.- Elliptic polarization

It can be shown (refs. 19 and 20) that an antenna which transmits elliptically polarized waves can be used to receive energy from an incident elliptically polarized wave. If the antenna polarization exactly matches the wave polarization, maximum energy is extracted from the passing wave. If not, there is an effective power loss or reflection. The polarization power loss factor, which is essentially an efficiency factor, is given as

$$K_p = \frac{1}{2}\left[1 + \frac{\pm 4 A_T A_R + \left(1 - A_T^2\right)\left(1 - A_R^2\right)\cos 2\alpha}{\left(1 + A_T^2\right)\left(1 + A_R^2\right)}\right] \quad (1)$$

where

A_T = axial ratio of the incident wave

A_R = axial ratio of the receiving antenna

α = angle between major axes of the polarization ellipses

+ = used for same sense rotation

− = used for opposite sense rotation

$K_p = 1$ indicates maximum power extracted from the wave. $K_p = 0$ indicates no power extracted from the wave. It can be easily shown that

$K_p = 1$ for circular polarization with the same sense or linear polarization with parallel axes. (2)

H-18

$$K_p = \frac{1}{2} \quad \text{for circular to linear polarization.} \tag{3}$$

$$K_p = 0 \quad \text{for circular polarization with opposite sense or linear polarization with perpendicular axes.} \tag{4}$$

H.6 Intelligibility of Clipped Voice

Studies of the intelligibility of peak clipped voice waveforms as a function of clipping depth have been performed by several authors and agencies. The results of investigations by Shyne (ref. 21) and Licklider (ref. 22) have been summarized and manipulated by Kadar of Grumman Aircraft and Engineering Corporation (ref. 23). It is not intended to resummarize here the results of the references. Rather, the applicable results from the references will be stated.

Plots of empirically derived data relating percent intelligibility for single words and percent word articulation to post detection peak speech to root-mean-squared noise, with peak clipping depth as a parameter, have been plotted in the references. The ordinates of the plots are linear in percent intelligibility. The abscissas are linear in decibels, peak speech to r.m.s. noise. Since the decibel value of a peak to r.m.s. ratio is the same as that of a peak-squared to mean-squared ratio, and since the ratio of peak-squared signal to mean-squared noise has been derived for several detectors in appendices D and E, the intelligibility plots are directly useable in predicting the performance of voice channels.

Manned Spacecraft Center
    National Aeronautics and Space Administration
        Houston, Texas, December 30, 1965

REFERENCES

1. Giacoletto, L. J.: Generalized Theory of Multitone Amplitude and Frequency Modulation. Proc. IRE, July 1947, p. 680.

2. Bennett, William R.: Electrical Noise. McGraw-Hill Book Co., Inc., 1960, p. 234.

3. Davenport, Wilbur B., Jr.; and Root, William L.: An Introduction to the Theory of Random Signals and Noise. McGraw-Hill Book Co., Inc., 1958, p. 158.

4. Davenport, W. B., Jr.: Signal-to-Noise Ratios in Band-Pass Limiters. Journal of Applied Physics, vol. 24, no. 6, June 1953, p. 720.

5. Middleton, David: An Introduction to Statistical Communication Theory. McGraw-Hill Book Co., Inc., 1960, p. 669.

6. Kuo, Benjamin C.: Automatic Control Systems. Prentice-Hall, Inc., 1962, p. 123.

7. Martin, Benn D.: The Pioneer IV Lunar Probe: A Minimum-Power FM/PM System Design. Tech. Rept. 32-215, Jet Propulsion Laboratory, Pasadena, Calif., March 15, 1962, p. 14.

8. Develet, Jean A. Jr.: An Analytic Approximation of Phase-Lock Receiver Threshold. Space Technology Laboratories, Inc., Redondo Beach, Calif., Apr. 10, 1962.

9. Korn, Granino A.; and Korn, Theresa M.: Mathematical Handbook for Scientists and Engineers. McGraw-Hill Book Co., Inc., 1961, p. 18.8-3.

10. Titsworth, R. C.; and Welch, L. R.: Power Spectra of Signals Modulated by Random and Pseudo Random Sequences. Tech. Rept. 32-140, Jet Propulsion Laboratory, Pasadena, Calif., Oct. 10, 1961, p. 21.

11. Martin, Jan W.: Apollo Pseudo-Random Noise Ranging System. To be published as a NASA TN in 1965.

12. Easterling, M.; and Schottler, P. H: Maximum Likelihood Acquisition of a Ranging Code. Research Summary 36-12, vol. I, Jet Propulsion Laboratory, Pasadena, Calif., Jan 2, 1962, p. 81.

13. Painter, J. H.; and Hondros, G.: Unified S-Band Telecommunications Techniques for Apollo - Volume I. NASA TN D-2208, 1965.

14. Middleton, David: An Introduction to Statistical Communication Theory. McGraw-Hill Book Co., Inc., 1960, p. 606.

15. Papoulis, Athanansios: The Fourier Integral and its Applications. McGraw-Hill Book Co., Inc., 1962, p. 247.

16. Livingston, Marvin L.: The Effect of Antenna Characteristics on Antenna Noise Temperature and System SNR. IRE Transactions on Space Electronics and Telemetry, Sept. 1961, p. 71.

17. Korn, Granino A.; and Korn, Theresa M.: Mathematical Handbook for Scientists and Engineers. McGraw-Hill Book Co., Inc., 1961, p. 9.3-10.

18. Schelkunoff, Sergie A.; and Friis, Harold T.: Antennas, Theory and Practice. John Wiley and Sons, Inc., New York, 1952, p. 43.

19. Sichak and Milazzo, S.: Antennas for Circular Polarization. Proc. IRE, vol. 36, Aug. 1948, p. 997.

20. Rumsey, V. H.; et al: Techniques for Handling Elliptically Polarized Waves, with Special Reference to Antennas. Proc. IRE, vol. 39, May 1951, p. 533.

21. Shyne, N. A.: Speech Signal Processing and Applications to Single Sideband. ERL Tech. Rept., Montana State College, Boseman, Mont., 1962.

22. Licklider, J. C. R.; and Pollack, I.: Effects of Differentiation, Integration, and Infinite Peak Clipping Upon the Intelligibility of Speech. Journal of the Acoustical Society of America, no. 20, 1948.

23. Kadar, I.: Speech Intelligibility Criteria for Apollo VHF AM Communication Links. LED-380-4, Grumman Aircraft Engineering Corp., Bethpage, L.I., Oct. 18, 1963.

"The aeronautical and space activities of the United States shall be conducted so as to contribute . . . to the expansion of human knowledge of phenomena in the atmosphere and space. The Administration shall provide for the widest practicable and appropriate dissemination of information concerning its activities and the results thereof."

—NATIONAL AERONAUTICS AND SPACE ACT OF 1958

NASA SCIENTIFIC AND TECHNICAL PUBLICATIONS

TECHNICAL REPORTS: Scientific and technical information considered important, complete, and a lasting contribution to existing knowledge.

TECHNICAL NOTES: Information less broad in scope but nevertheless of importance as a contribution to existing knowledge.

TECHNICAL MEMORANDUMS: Information receiving limited distribution because of preliminary data, security classification, or other reasons.

CONTRACTOR REPORTS: Technical information generated in connection with a NASA contract or grant and released under NASA auspices.

TECHNICAL TRANSLATIONS: Information published in a foreign language considered to merit NASA distribution in English.

SPECIAL PUBLICATIONS: Information derived from or of value to NASA activities. Publications include conference proceedings, monographs, data compilations, handbooks, sourcebooks, and special bibliographies.

TECHNOLOGY UTILIZATION PUBLICATIONS: Information on technology used by NASA that may be of particular interest in commercial and other nonaerospace applications. Publications include Tech Briefs; Technology Utilization Reports and Notes; and Technology Surveys.

Details on the availability of these publications may be obtained from:

SCIENTIFIC AND TECHNICAL INFORMATION DIVISION

NATIONAL AERONAUTICS AND SPACE ADMINISTRATION

Washington, D.C. 20546

www.ingramcontent.com/pod-product-compliance
Lightning Source LLC
Chambersburg PA
CBHW081723170526
45167CB00009B/3681